STUDIES IN FUNERARY ARCHAEOLOGY 20

Ethnoarchaeology of Rock-cut Tombs

A Study of Toraja Cemeteries (Sulawesi, Indonesia)

Guillaume Robin and Ron Adams

Oxford & Philadelphia

Published in the United Kingdom in 2025 by
OXBOW BOOKS
81 St Clements, Oxford OX4 1AW

and in the United States by
OXBOW BOOKS
1950 Lawrence Road, Havertown, PA 19083

© Oxbow Books and the individual contributors 2025

Paperback Edition: ISBN 979-8-88857-220-7
Digital Edition: ISBN 979-8-88857-221-4

A CIP record for this book is available from the British Library

Library of Congress Control Number: 2025938880

All rights reserved. No part of this book may be reproduced or transmitted in any form or by any means, electronic or mechanical including photocopying, recording or by any information storage and retrieval system, without permission from the publisher in writing.

Printed in Malta by Melita Press
Typeset in India by DiTech Publishing Services

For a complete list of Oxbow titles, please contact:

UNITED KINGDOM	UNITED STATES OF AMERICA
Oxbow Books	Oxbow Books
Telephone (0)1226 734350	Telephone (610) 853-9131, Fax (610) 853-9146
Email: oxbow@oxbowbooks.com	Email: queries@casemateacademic.com
www.oxbowbooks.com	www.casemateacademic.com/oxbow

Oxbow Books is part of the Casemate Group

Front cover: Rock-cut chambered tomb in the process of being cut by specialised stone artisans in Lemo (Makale Utara, Tana Toraja) (Photo: Guillaume Robin, June 2017).
Back cover: Rock-cut tomb cemetery of Lemo (Makale Utara, Tana Toraja) (Photo: Guillaume Robin, June 2017).

The Publisher's authorised representative in the EU for product safety is Authorised Rep Compliance Ltd., Ground Floor, 71 Lower Baggot Street, Dublin D02 P593, Ireland.
www.arccompliance.com

Contents

List of figures, plates and tables v

1. Introduction 1
   Rock-cut monuments and society 1
   Toraja rock-cut tombs (*liang pa'*) in a nutshell 2
   The physical environment: Toraja geography and geology 4
   The social environment: Toraja society between tradition and modernity 8
   Toraja funeral ceremonies 13
   Studies on Toraja stone monuments: from ethnography to ethnoarchaeology 16
   Fieldwork 2017 18
   Acknowledgments 20

2. Tomb traditions through time and space in Tana Toraja 21
   Wooden sarcophagi (*erong*) 21
   Rock-cut tombs (*liang pa'*) 35
   House-tombs (*patane*) 42
   Lower rank burials 50
   Infant burials 50

3. The anatomy and decoration of *liang pa'* 53
   Burial chamber 53
   Entrance area 59
   Closing systems 61
   Decorations 63
   Conclusion 75

4. Creating a *liang pa'*: cutting process and rituals 78
   When and where are new *liang pa'* created? 79
   The stone workers: costs, workspaces, tools and roles 83
   Cutting out the tomb: a technical and ritual *chaîne-opératoire* 86
   Consecration of the tomb 88
   Cutting failures and abandonments 92
   Tomb worksites as stone quarries 93
   Conclusion 93

5. Using a *liang pa'*: burial and post-burial rituals 96
   Body preparations prior to burial 96
   Ceremonial artefacts: from production to procession and deposition
   at the tomb 100
   Opening of the tomb and entombment 108
   Post-burial depositions outside the tomb 115
   Ma'nene' ritual 117
   Relocation of burials 119
   Conclusion 119

6. The landscape setting of *liang pa'* cemeteries 121
   Stone and death: Toraja cemeteries as rocky places 121
   Who owns the cemeteries? Communal *vs* private lands 122
   The importance of cardinal points in Toraja cosmography
   and ritual practices 124
   Landscape relationships between cemeteries and villages 125
   Landscape orientation of the tombs 127
   Boulder cemeteries: case studies 131
   Conclusion 142

7. The social biography of *liang pa'* cemeteries 147
   Mapping kinship groupings: the geographical catchment of cemeteries 147
   The social life of cemeteries 151
   Conclusion 153

8. Conclusion 155
   The social value of rock-cut tombs 155
   The architecture and decoration of rock-cut tombs 156
   The construction of rock-cut tombs 157
   Burial practices 158
   The landscape setting of rock-cut tombs 159
   Cemeteries and kinship systems 160
   Impact of colonial and post-colonial changes on rock-cut tomb practices 161

Glossary of Toraja terms 164
Appendix: list of cemeteries surveyed in June 2017 167
Bibliography 177

List of figures, plates and tables

Figures

Figure 1.1: Photogrammetry survey of a recently completed rock-cut tomb in Suloara' (Sesean Suloara'). 3
Figure 1.2: Cliff cemetery at Lemo (Makale Utara), with rectangular entrances to rock-cut tombs and *tau-tau* human effigies of the dead in rock-cut galleries. 3
Figure 1.3: Location of Tana Toraja on the island of Sulawesi, Indonesia. 4
Figure 1.4: Administrative map of Sulawesi, showing the location of the six provinces. 5
Figure 1.5: Topographic and administrative maps of the Toraja country and districts. 6
Figure 1.6: Rock-cut tomb cemetery created in a group of basalt boulders in Batutumonga (Sesean Suloara'). 7
Figure 1.7: Cemetery of Suaya (Sangalla), cut into a limestone cliff of the Makale formation. 9
Figure 1.8: Traditional *tondok* (hamlet) of Karuaya (Sangalla Utara), with its row of *tongkonan* (kinship houses) on the right, facing a row of *alang* (rice granaries) on the left. 10
Figure 1.9: Front of a *tongkonan* (kinship house) at Karuaya, with stacks of buffalo horns resulting from sacrifices at past funeral ceremonies. 11
Figure 1.10: Erection of a standing stone at a *rante* (ritual plaza) in Parinding (Sesean) as part of a funeral ceremony. 12
Figure 1.11: House-like *saringan* (palanquin) with a wrapped body ready to be carried outside of the *tondok* of Tandung La'bo (Sanggalangi) as part of a high-ranked funeral ceremony. 15
Figure 2.1: Two main styles of *erong* (wooden sarcophagi) from the Sa'dan region of Tana Toraja. 23
Figure 2.2: Location of *erong* cemeteries in the Sa'dan region of Tana Toraja. 24
Figure 2.3: *Erong* sarcophagi placed on wooden support inserted into the limestone cliff of Pala'tokke (Sanggalangi). 25
Figure 2.4: Rock-shelter of Marante (Tondon), which was used as an *erong* cemetery. 26
Figure 2.5: Rock-shelter *erong* cemetery of Lombok Parinding (Sesean), with stacks of old abandoned sarcophagi. 27

Figure 2.6: House-shaped *erong* at Tampangallo (Sangalla) and buffalo-shaped *erong* at Lombok Parinding (Sesean). 28

Figure 2.7: Incised ornamentation on the side of an *erong* sarcophagus at Tampangallo (Sangalla). 29

Figure 2.8: *Erong* sarcophagi with incised ornamentation at Lombok Parinding (Sesean) and Ke'te' Kesu (Kesu). 30

Figure 2.9: *Erong* sarcophagi with incised ornamentation at Tampangallo (Sangalla) and Ke'te' Kesu (Kesu). 31

Figure 2.10: *Pa'tedong* (buffalo head) motif incised among geometric motifs on an *erong* sarcophagus at Lombok Parinding (Sesean). 32

Figure 2.11: *Erong* sarcophagus with *pa'sussu'* (vertical grooves) ornamentation at Tampangallo (Sangalla) and *pa'sussu'* decoration on a rice barn at Karuaya (Sangalla Utara). 33

Figure 2.12: Coffin of a recent aristocratic burial placed at a very high cliff crevice at Londa (Kesu). 35

Figure 2.13: Rock-cut tomb cemetery of Lo'ko' Mata in Tonga Riu (Sesean Matallo) in 1979 and 2017. 36

Figure 2.14: A very old *liang pa'* rock-cut tomb (Parinding boulder 10, Sesean). 38

Figure 2.15: Two generations of *liang pa'* rock-cut tombs on boulder 21 at Bori' (Sesean). 39

Figure 2.16: Location of rock-cut tomb cemeteries examined by the authors in 2017 in the Sa'dan region of Tana Toraja. 42

Figure 2.17: Location of rock-cut tomb cemeteries examined by the authors in 2017 in the Sesean districts (Area A). 43

Figure 2.18: Development over time of the three main types of aristocratic burial in Tana Toraja. 44

Figure 2.19: Small-scale wooden house replica on top of a *patane* tomb dug into a flat outcrop in the rice paddies of Kalambe (Tikala) c. 1930. 44

Figure 2.20: The old wooden *patane* of Pangkaro in Tembamba (Buntao'), today protected under a corrugated roof shelter. 46

Figure 2.21: A traditional wooden *patane* of recent facture in the cemetery of Buntu Pune (Kesu). 47

Figure 2.22: Modern concrete *patane* in La'bo' (Sanggalangi). 48

Figure 2.23: *Liang piah* infant burials in Kambira (Sangalla). 51

Figure 3.1: Western face of boulder 4 in Deri (Sesean), with an old *liang pa'* (style 1) and a new *liang pa'* (style 2). 54

Figure 3.2: A newly hewn *liang pa'* at Lemo (Makale Utara). The unusual length of the entrance area is due to the presence of a rock fault which the masons had to avoid to create the burial chamber. 57

List of figures, plates and tables vii

Figure 3.3: Boulder 26 at Lempo (Sesean Suloara'), with its recently hewn *liang pa'*. The tomb's chamber has a separate recess space which is reserved for the coffin of the sponsor of the tomb. 58

Figure 3.4: Boulder 1 at Bori' (Sesean), with ancient *liang pa'* rock-cut tombs. 60

Figure 3.5: Buffalo head motifs (*pa'tedong*) carved on the wooden shutter of ancient *liang pa'* rock-cut tombs. 61

Figure 3.6: Basket lid motif (*pa'kapu baka*) carved on the wooden shutter of an ancient *liang pa'* and ancient *liang pa'* with geometrical *pa'erong* motifs on the jamb and sill panels. 62

Figure 3.7: Tomb doors with sculpted human representations (*tau-tau*). 63

Figure 3.8: Entrances of recent *liang pa'* rock-cut tombs with wood-carved door shutters. 67

Figure 3.9: Undecorated tomb door (Buntu Lobo boulder 12, Sesean) and Christian cross painted on a tomb door (Batutumonga boulder 9, Sesean Suloara'). 68

Figure 3.10: Rock-cut tombs with stone-made door shutters. 69

Figure 3.11: Buffalo-head reliefs (*kabongo'*) sculpted into the rock under the entrances of rock-cut tombs. 71

Figure 3.12: Buffalo-head reliefs (*kabongo'*) sculpted into the rock under the entrances of rock-cut tombs. 72

Figure 3.13: A richly decorated tomb on boulder 7 at Lempo (Sesean Suloara'), with rock-carved *kabongo'* (buffalo-head sculpture), *pa'barana'* (leaves), *pa'bulu londong* (feathers) motifs, surrounding a wooden door adorned with the *pa'barre allo* (sun) motif. 73

Figure 3.14: Rock-carved *pa'barana'* (leaves) and *pa'bulu londong* (feathers) motifs surrounding the entrance of a tomb on Lempo boulder 41 (Sesean Suloara'), with human face and upper torso. 75

Figure 3.15: Rock-carving surrounding the entrance of a tomb in boulder 14 in Lempo (Sesean Suloara'). The two human figures on each side of the entrance represent the couple who sponsored the cutting of the tomb. 76

Figure 3.16: Rare occurrence of rock-carved *sembang* (house-beams) on each side of a tomb entrance in Bori' boulder 10 (Sesean). 77

Figure 4.1: *Liang pa'* rock-cut tomb in the process of being cut at Lemo (Makale Utara) in June 2017. 80

Figure 4.2: Stone workers cutting a *liang pa'* rock-cut tomb around 1935 at an unknown location in Tana Toraja. 81

Figure 4.3: Carved and painted rectangular marks used to reserve locations for future rock-cut tombs on cemetery rock faces. 82

Figure 4.4: A *liang pa'* worksite at Suloara' boulder 10 (Sesean Suloara'). 85

Figure 4.5:	Metal picks and hammer used by stone workers to hew out rock-cut tombs.	86
Figure 4.6:	Stone workers working in rotation.	86
Figure 4.7:	Sequence of cutting work for creating a *liang pa'* rock-cut tomb.	89
Figure 4.8:	Cutting and flattening the floor of the chamber is the last part of the hewing process.	90
Figure 4.9:	Installation of a wooden door on a rock-cut tomb in Lempo boulder 18 (Sesean Suloara').	91
Figure 4.10:	A recently completed *liang pa'* rock-cut tomb in Lempo (Sesean Suloara'). It will be left open until its consecration.	92
Figure 4.11:	Woman collecting stone chips at a *liang pa'* worksite in Bori' (Sesean) and field retaining wall in Suloara' (Sesean Suloara'), built from stone blocks that were extracted during the cutting of the tomb visible in the background.	94
Figure 5.1:	Funeral procession in Lempo (Sesean Suloara'): a deceased individual is carried across rice paddies by relatives in a *saringan* palanquin to its rock-cut tomb.	101
Figure 5.2:	*Tau-tau* effigy of Ne' Lai' Ambun, placed under a rice barn as part of her high-ranked funeral ceremony in 1937 in Tadongkon (Kesu).	104
Figure 5.3:	Rock-cut tombs and *tau-tau* effigies in rock-cut galleries photographed by Albert Grubauer in 1911 in Tondong Dandelolo near Makale.	105
Figure 5.4:	*Tau-tau* effigies in a rock-cut gallery at the cemetery of Suaya (Sangalla).	106
Figure 5.5:	A *dulang* wooden bowl in a private collection in Tonga Riu and dulang bowls deposited under the entrances of ancient *liang pa'* rock-cut tombs in Parinding boulder 9 (Sesean).	108
Figure 5.6:	Placement of a wrapped body inside a *liang pa'* rock-cut tomb in Lempo boulder 18 (Sesean Suloara').	111
Figure 5.7:	Burials inside old *liang pa'*, whose door shutters are missing.	112
Figure 5.8:	*Liang pa'* in Bori' boulder 26 (Sesean) with an internal wooden wall dividing the chamber into two spaces for the relatives of two individuals.	115
Figure 6.1:	Orientations of the 697 *liang pa'* rock-cut tombs surveyed in June 2017, in relation to their landform context.	129
Figure 6.2:	Lo'ko' Mata boulder in Tonga Riu (Sesean Suloara'), with its 95 rock-cut tombs distributed on multiple faces.	132
Figure 6.3:	Boulder 10 in Parinding (Sesean). This large, pedestalled erratic was used as a rock-shelter for *erong* sarcophagi and subsequently hewn out with six *liang pa'* rock-cut tombs.	134

Figure 6.4:	Boulder 7 in Parinding (Sesean).	135
Figure 6.5:	Plan and section of boulder 9 in Parinding (Sesean) showing locations of old and recent *liang pa'* rock-cut tombs.	137
Figure 6.6:	Boulder 1 in Bori' (Sesean) seen from east (left) and southwest (right).	138
Figure 6.7:	Boulder 4 in Bori' (Sesean) seen from southwest (top) and southeast (bottom).	140
Figure 6.8:	Boulder 21 in Bori' (Sesean) seen from the south.	141
Figure 6.9:	Plan of boulders 9 and 10 in Buntu Lobo (Sesean) showing locations of old and recent *liang pa'* rock-cut tombs.	141
Figure 6.10:	Map of boulders 3–7 and surroundings in Suloara' (Sesean Suloara') showing locations of old and recent *liang pa'* rock-cut tombs as well as those which were in the process of being cut in 2017.	143
Figure 6.11:	Map of boulders 1–8 and surroundings in Deri (Sesean) showing locations of old and recent *liang pa'* rock-cut tombs as well as those which were in the process of being cut in 2017.	144
Figure 6.12:	Diagram showing the differences in the location of old and recent *liang pa'* rock-cut tombs in the landscape and in relation to settlements.	145
Figure 7.1:	*Tongkonan* Papa Batu ('stone roof') in Banga (Rembon) with its unique stone-tiled roof.	150
Figure 7.2:	The old, disused cemetery of Sele in Suloara' (Sesean Suloara').	152

Plates

Plate 1:	Cliff cemetery at Lemo (Makale Utara), with rectangular entrances to rock-cut tombs and *tau-tau* human effigies of the dead in rock-cut galleries.
Plate 2:	Topographic map of Sulawesi, with a close-up view of the Tana Toraja area.
Plate 3:	Geological map of Sulawesi.
Plate 4:	Volcanic landscape on the slope of Mount Sesean, with basalt outcrops and boulders.
Plate 5:	Main decorative motifs used on the wooden doors of *liang pa'* rock-cut tombs.
Plate 6:	Recent *liang pa'* rock-cut tombs with decorated wooden doors on Bori' boulder 21 (Sesean). The two lower tombs have a *pa'tedong* (buffalo head) motif, while the two upper ones have a *pa'barre allo* (sun) motif.
Plate 7:	Anatomy of older *liang pa'* rock-cut tombs (style 1).
Plate 8:	Entombment at Lempo boulder 18 (Sesean Suloara'): the deceased's body, wrapped in a cylinder of cloths, is carried down the rock to the entrance of the rock-cut tomb.

Plate 9: Location of To Bua' cemetery (Deri) in relation to local *tongkonan* houses and other boulder cemeteries.
Plate 10: Satellite view of Ke'te' Kesu (Kesu), with locations of its *tondok* (village), *rante* (ceremonial plaza) and cliff cemetery.
Plate 11: Satellite view of Buntu Pune (Kesu), with locations of its cliff cemetery and ancient and current *tondok* (village).
Plate 12: Distribution of boulders with *liang pa'* rock-cut tombs and *erong* sarcophagi in Parinding (Sesean).
Plate 13: Distribution of boulders with *liang pa'* rock-cut tombs in Bori' (Sesean).
Plate 14: Distribution of boulders with *liang pa'* rock-cut tombs in Buntu Lobo (Sesean).
Plate 15: Rock-cut tomb cemetery of Batu Lappa' (Buri', Rembon) (photo: G. Robin).
Plate 16: Location of *tongkonan* kinship houses associated with the rock-cut tomb cemeteries of Batu Lappa' and Sanduni' (Rembon).
Plate 17: Rock-cut tomb cemetery of Sele in Suolara' (Sesean Suloara').
Plate 18: Tongkonan kinship houses associated with the cemetery of Sele.

Tables

Table 2.1: Proportion of old and recent *liang pa'* rock-cut tombs in study area A. 41
Table 2.2: Main differences between ancient and recent *patane* house-tombs. 49
Table 3.1: Measurements from 21 recent *liang pa'* (style 2). 56
Table 3.2: Preservation of doors and decorations in old (style 1) *liang pa'* rock-cut tombs. 64
Table 3.3: Preservation of doors and decorations in recent (style 2) *liang pa'* rock-cut tombs. 65
Table 3.4: List of tombs with rock-carved motifs in the Sesean districts (study area A). 74
Table 4.1: Data collected at ten rock-cut tombs in process of being cut in June 2017. 84
Table 4.2: Names of the main steps required for creating a *liang pa'* rock-cut tomb, with their associated stone cutting and ritual activities. 87
Table 6.1: Orientations of the 697 *liang pa'* rock-cut tombs surveyed in June 2017, in relation to their landform context. 128
Table 6.2: Orientations of the 697 *liang pa'* rock-cut tombs surveyed in June 2017, in relation to their date. 129
Table 6.3: Orientations of the 697 *liang pa'* rock-cut tombs surveyed in June 2017, in relation to slope aspect of their landscape setting. 130
Table 6.4: Proportions of boulders from Mount Sesean (study area A) having *liang pa'* rock-cut tombs in one or more faces. 131

Table 6.5:	Orientation of old and recent *liang pa'* on the boulder cemetery of Lo'ko' Mata in Tonga Riu (Sesean Suloara').	133
Table 6.6:	Orientation of old and recent *liang pa'* on Parinding boulder 4 (Sesean).	136
Table 6.7:	Orientation of old and recent *liang pa'* on Parinding boulder 9 (Sesean).	136
Table 6.8:	Orientation of old and recent *liang pa'* on Bori' boulder 1 (Sesean).	138
Table 6.9:	Orientation of old and recent *liang pa'* on Bori' boulder 21 (Sesean).	139
Table 6.10:	Main differences in the landscape location of old and recent *liang pa'* rock-cut tombs.	145

Chapter 1

Introduction

Rock-cut monuments and society

Social interactions with the mineral world and underground spaces are some of the most prevalent human activities around the globe, from cave explorations and rituals, to mining and quarrying outcrops. Rock-cut monuments are part of this universal human experience with the physical environment. Unlike stone-built monuments, rock-cut monuments are directly inserted into the landscape. They require a significant alteration of, and social investment in, a rocky place for the creation of an architectural space. Importantly, this alteration is irreversible and durable. Rock-cut monuments, therefore, provide societies with powerful tools for strategies of commemoration, land ownership and identity expression. It is not surprising that many societies have adopted them, from prehistoric hypogea chambered tombs in the Mediterranean (Melis 2000), Japan, Korea (Ikegami 2018), and the Americas (Sevilla Casas 2010), to Nabataean temples and tombs in early historical Near East (McKenzie 1990), medieval rock-cut churches in Ethiopia (Muehlbauer 2023), and up to modern times with massive projects such as the Mount Rushmore sculptures in the USA.

Rock-cut architectures are not just buildings, they are social appropriations of the mineral environment. As products of cultural constructs, they are indicative of societies' social organisation and belief systems. Their architecture and decoration are designed to serve particular social needs that are locally specific. This is why the character of rock-cut monument traditions differs so much across space and time (see examples above). Rock-cut architectures are also indicative of cultural perceptions of the environment. They are not built anywhere in the landscape; their location is determined not only by geological factors but also by political and cosmological considerations. As permanent, visible monuments, they may serve as territorial markers and/or to enhance places that are invested with particular beliefs or powers. For these different reasons, rock-cut monuments are key entries into the societies who build them. Moreover, they provide us with fascinating insights into human perceptions and interactions with the mineral world, which have implications beyond the chosen case study period or area. This is why rock-cut architectures are worthy of scholarly enquiry.

The vast majority of rock-cut monument traditions are relics of the past and can only be studied archaeologically. Only very few regions have maintained living traditions of hewing practices associated with ritual activities. For instance, rock-cut churches are still being cut in Ethiopia (Fauvelle-Aymar *et al.* 2010; Lamesa *et al.* 2023). Sulawesi, Indonesia, is the only place in the world where rock-cut tombs are still being made today. These burial monuments, locally called *liang pa'* ('cut tombs'), are specific to the Toraja culture of South Sulawesi. This tradition has been impacted by societal changes in recent decades and may eventually cease to exist. This is why it is crucial to document and study the tradition before the invaluable cultural knowledge associated with it disappears. As such, it is also important to present this unique tradition to both specialist and non-specialist audiences alike.

This book is the first ever dedicated to Toraja *liang pa'*. Its aim is twofold: first, to provide an overview of *liang pa'* rock-cut tombs, based on a review of ethnographic literature and on field observations; second, to address specific issues that had never been investigated before. These issues include the architectural style and decoration of the tombs (Chapter 3); the technical steps and ritual activities associated with the process of cutting them into the rock (Chapter 4); their landscape setting (Chapter 6); and their relationship to local kinship groups (Chapter 7). Chapter 2 places the *liang pa'* within the development of other funerary traditions of the Toraja people in the last centuries, while Chapter 5 presents an overview of burial practices associated with these monuments. Before we tackle those core themes, the current chapter introduces the reader to our topic and its context.

Toraja rock-cut tombs (*liang pa'*) in a nutshell

Liang is the general Toraja term for 'tomb' or 'grave' (Grubauer 1913; Keers 1939; Waterson 1988). It is commonly used to refer to different types of tombs used by Toraja people, including rock-cut tombs. When referring specifically to rock-cut tombs, as opposed to other types of tombs, Toraja people use the term *liang pa'* ('cut tomb') (Brisbois and Douvier 1980; Duli 2018; Duli *et al.* 2019) or, less frequently, *liang batu* ('stone tomb') (Koubi 1982, 196). Toraja rock-cut tombs are simple architectures. They consist of a single rectangular chamber closed by a small wooden door (Fig. 1.1). The chamber is just deep enough to receive bodies in extended (supine) position (*i.e.* 1.8–2.0 m depth). The width and height of chambers vary from one tomb to another but normally range from 1–2 m. *Liang pa'* are generally grouped together on large vertical rock faces, where they form cemeteries (Fig. 1.2; Pl. 1), although isolated tombs are sometimes found. Some tombs are made on high cliffs and are only accessible by long bamboo ladders. Others, especially the ones made in recent years, are located closer to the ground.

1. Introduction

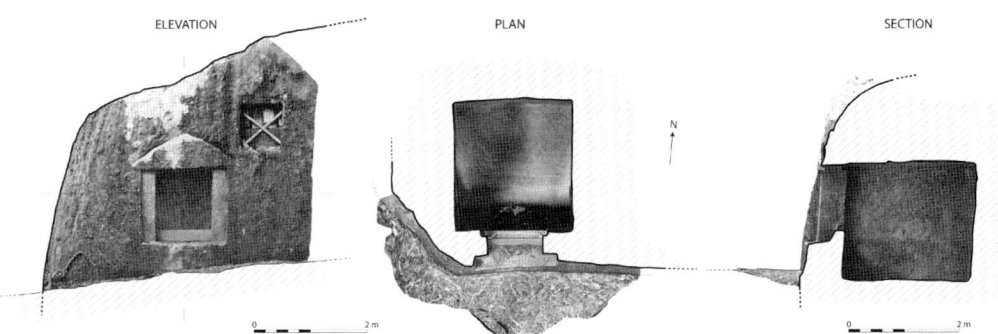

Figure 1.1: Photogrammetry survey of a recently completed rock-cut tomb in Suloara' (Sesean Suloara'). The wooden door (missing here) is normally added just before the first interment takes place (images: G. Robin).

Figure 1.2: Cliff cemetery at Lemo (Makale Utara), with rectangular entrances to rock-cut tombs and tau-tau *human effigies of the dead in rock-cut galleries (photo: G. Robin).*

The tradition likely started in the 17th century AD, and has been maintained since then, with specialist stone carvers still very active today. Rock-cut tombs are not found everywhere in Toraja country. They are specific to the Sa'dan region, *i.e.*, the catchment basin of the river Sa'dan, from its source in Mount Sesean to the districts around the city of Makale (see Fig. 1.5). Rock-cut tombs are absent in areas west of the river Masuppu, where large rock outcrops are less frequent. No other groups in Sulawesi or elsewhere in Indonesia use rock-cut tombs for their dead. It is therefore an exclusively Sa'dan Toraja tradition.

Liang pa' are collective tombs. They are used by families, sometimes over many generations. Each family has its own tomb. They are also aristocratic tombs. In the past, only noble individuals of high ranks could create and use *liang pa'*, while lower rank nobles and commoners were buried in cliff crevices and caves. Nowadays, however, anybody with a link to a noble family and enough wealth can sponsor the creation of a rock-cut tomb for themselves and their descendants. Despite this actual democratisation, the social value of *liang pa'* has not diminished and is still associated with high rank and prestige in Tana Toraja.

The *liang pa'* tradition is the coalescence of two main elements: the Toraja culture (and its focus on the ancestors) and the geological landforms of South Sulawesi. We will now turn to these two elements and look at the main characteristics of the social and environmental contexts within which Toraja rock-cut tombs are created and used.

The physical environment: Toraja geography and geology

Tana Toraja is the name of the traditional territory of the Toraja people. It is located in the southern part of the island of Sulawesi in Indonesia (Figs 1.3; 1.4). The Toraja territory represents a surface of 3332.75 km². Administratively, the area is split

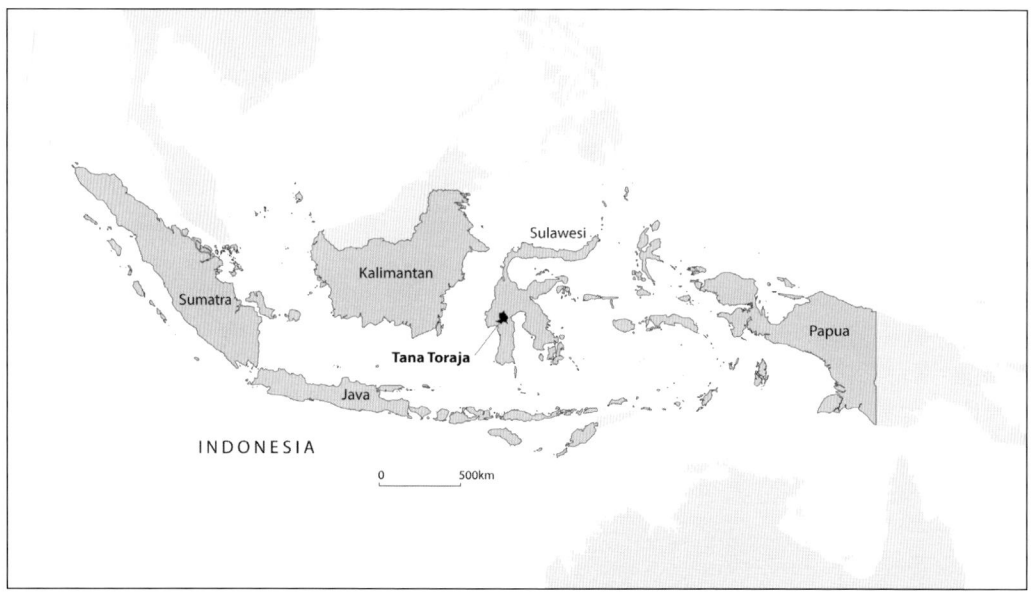

Figure 1.3: Location of Tana Toraja on the island of Sulawesi, Indonesia (map: G. Robin).

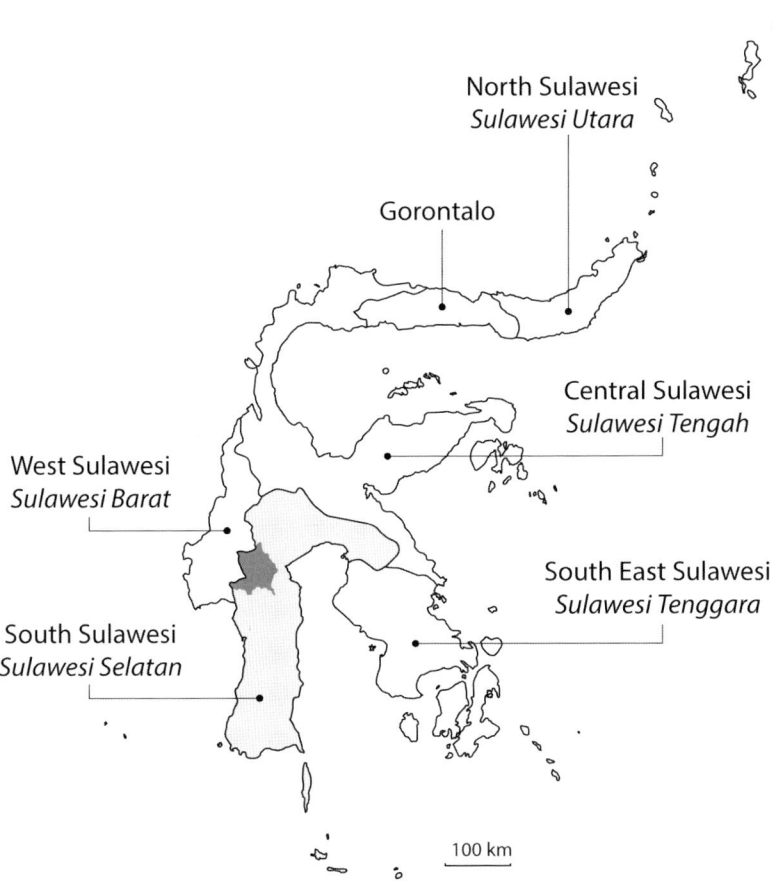

Figure 1.4: Administrative map of Sulawesi, showing the location of the six provinces. Tana Toraja is represented by the dark grey area in the South Sulawesi province (map: G. Robin).

into two regencies (Indonesian *kabupaten*): Tana Toraja ('Land of the Toraja') in the south, with the town of Makale as its capital, and Toraja Utara ('North Toraja') in the north, with Rantepao as its capital (Fig. 1.5). Before 2008, Toraja Utara and Tana Toraja formed a single regency, called Tana Toraja, with Rantepao as capital. In the literature, and in the present book, Tana Toraja is used to refer to the entire Toraja territory as opposed to the current southern regency only (unless specified otherwise).

Each regency is divided into *kecamatan* (districts). There are currently 40 districts across the two regencies: 21 in the northern regency (Toraja Utara) and 19 in the southern one (Tana Toraja) (Fig. 1.5). The number of districts and their boundaries have changed significantly over time, from 32 *kecamatan* during the Dutch administration,

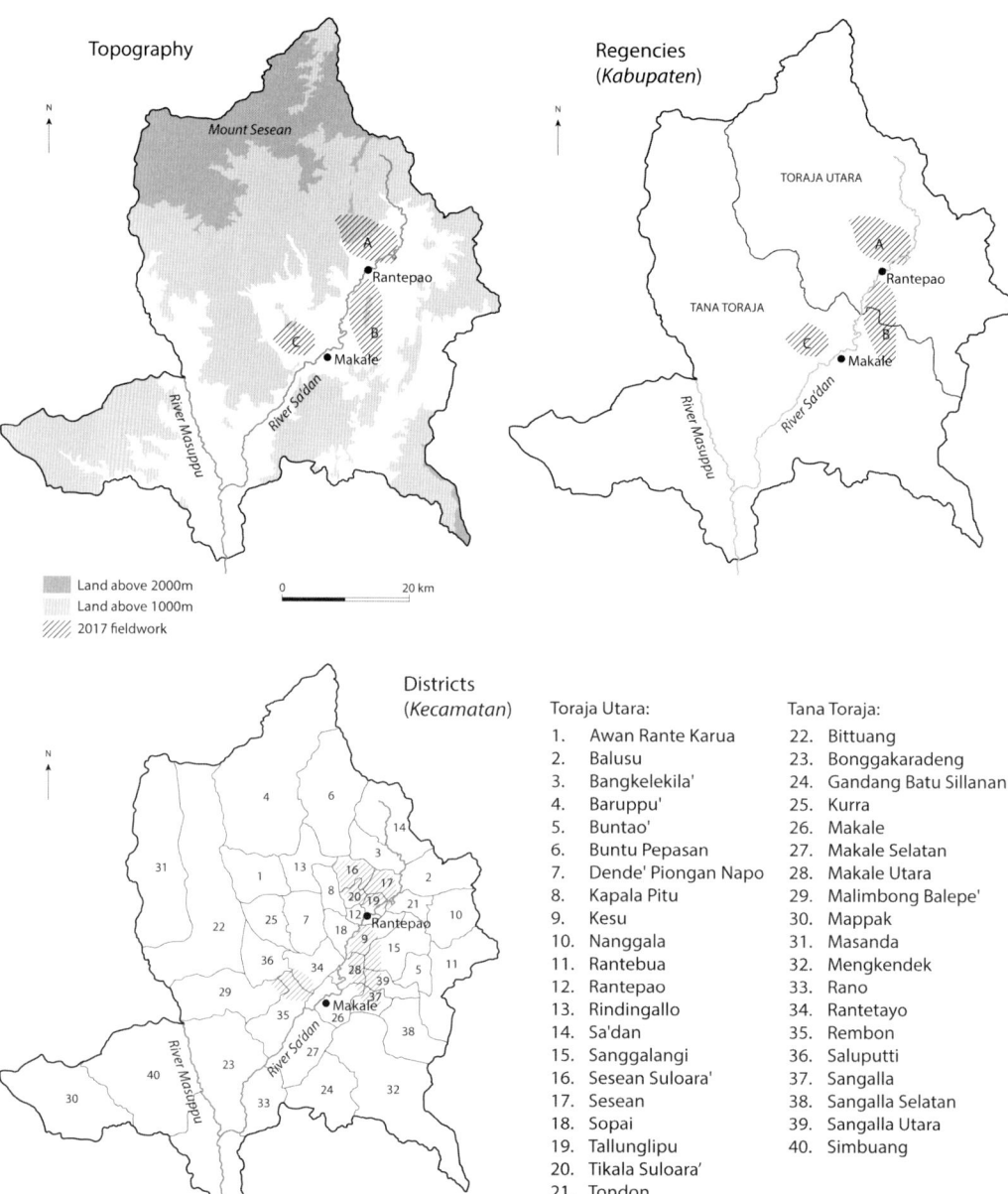

Figure 1.5: Top left: topographic map of the Toraja country (source: https://ngmdb.usgs.gov/topoview/viewer); top right: administrative map of the Toraja territory, with its two regencies (kabupaten) of Toraja Utara (north) and Tana Toraja (south); bottom: administrative map of the 40 districts (kecamatan) of the Toraja territory. Letters A, B, C refer to areas surveyed by the authors in 2017 (maps: G. Robin).

to nine in the 1970s, 15 in the 1980s, and 40 from 2008 (see Waterson 2009, xviii–xx, for maps of the first three stages). This sometimes creates confusion over the naming and localisation of places referred to in ethnographic sources, which were produced at different periods and therefore used different toponymic systems.

Tana Toraja is located in the internal, mountainous part of South Sulawesi (Fig. 1.5; Pl. 2). The nearest coastal areas are 20 km to the east, and altitudes range from 700 m in the Rantepao and Makale plains, to 2150 m above sea level at the peak of Mount Sesean in the north. Before Dutch colonisation in 1909, the area remained in relative isolation, politically independent from the neighbouring kingdoms located on the coasts of Sulawesi, such as the Bugis kingdom of Luwu. The word 'Toraja' may derive from the Buginese word *to ri aja*, meaning 'people of the uplands' (Waterson 2009, 9–10). The Toraja highlands are crossed by two main valleys in which flow the Sa'dan and the Massapu rivers (Fig. 1.5). This book will be more concerned with the Sa'dan region, where *liang pa'* are located.

Tana Toraja lies within a region with a dynamic geomorphological setting. Mountain ranges in the highlands of Tana Toraja formed as a result of volcanic activity that occurred between 150 and 15 million years ago. Most prominent among these mountains is Mount Sesean (2150 m) in the northwestern part of

Figure 1.6: Rock-cut tomb cemetery created in a group of basalt boulders in Batutumonga (Sesean Suloara') (photo: G. Robin).

Tana Toraja. Magmatic rocks associated with these ranges include basalt, rhyolites, and gabbros (Polvé *et al.* 1997, 83; White *et al.* 2017, 75). Makale Formation Reef Limestones from the lower to Middle Miocene geological epochs overlie the Lamasi complex rocks in many parts of southeastern and central Tana Toraja (White *et al.* 2017, 75) (Pl. 3).

As a consequence of these geological formations, the northwestern part of Tana Toraja tends to be marked by scattered outcropping basalts and other rocks of magmatic origin, often in the form of large erratic boulders (Fig. 1.6, Pl. 4). This contrasts with the southeastern and central parts of Tana Toraja where limestone cliffs, karstic caves and rock shelters are more dominant features. This marked differentiation in landforms has had an impact on the overall distribution and local configurations of *liang pa'* cemeteries. In the southeast, tombs are often found in high concentrations (*e.g.*, 10–70) within large, single limestone cliffs (Fig. 1.7), while in the northwest, tombs are often found isolated or in small groups (*e.g.*, 2–5) carved into individual igneous boulders or outcrops of various sizes that are scattered over hill slopes. Settlements and rice paddies are found both in valley floors and hill slopes (terraces) throughout Tana Toraja. Villages tend to be built on higher ground in relation to the rice fields.

The social environment: Toraja society between tradition and modernity

Tana Toraja is a culturally bounded area, located in a small highland area of the island of Sulawesi. It is culturally distinct from its neighbours within the island (and in Indonesia generally), in particular those on the coasts, such as the Bugis, who are associated with different historical trajectories and developed into kingdoms. Toraja society and cultural traditions are very rich and complex and will not be discussed in an exhaustive sense here (for a general reference see Waterson 2009). Instead, we will focus on key aspects of social organisation and belief systems that help illuminate the context and conditions under which *liang pa'* have been created and used.

Tana Toraja's highland landscape, marked by hills and valleys, has influenced mobility and settlement patterns. The country is traditionally comprised of numerous small villages or hamlets (*tondok*) that are scattered amongst large tracts of wet-rice paddy lands on valley floors and upland terraced configurations and connected by small paths and roads. The hamlets (often better described as household clusters) are typically composed of one or more kindred group houses (*tongkonan*) and rice granaries (*alang*) (Fig. 1.8). Each hamlet is surrounded by cultivated lands (*pateng*), woodlands (*pangala'*), a ceremonial plaza (*rante*), one or more cemeteries and other hamlets. The largest villages also each have a market area (Nooy-Palm 1979, 92).

The main agricultural activity is rice cultivation, although cash crops have more recently been introduced: first coffee in the 17th century, then cocoa and vanilla

Figure 1.7: Cemetery of Suaya (Sangalla), cut into a limestone cliff of the Makale formation (photo: G. Robin).

Figure 1.8: Traditional tondok *(hamlet) of Karuaya (Sangalla Utara), with its row of* tongkonan *(kinship houses) on the right, facing a row of* alang *(rice granaries) on the left (photo: G. Robin).*

in the 1970s–1980s (Nooy-Palm 1986, 304; Waterson 2009, 117–118; de Jong 2013). Animal husbandry is also widespread in the country, mainly with chicken, pigs and water buffaloes, the latter being used for social and ceremonial exchanges and sacrifices rather than as draft animals. The water buffalo is associated with wealth and status and is one of the main symbols of Toraja nobility. The animal is frequently depicted in decorative motifs on the wooden doors of houses and tombs (Robin 2017).

Toraja society was traditionally divided into three social classes: nobles, commoners (*to buda*, 'the many') and slaves (*kaunan*). The noble class is itself divided into two main hierarchical categories: upper nobles, called *tana' bulaan* ('golden stake'), or *puang* ('lords') in the southeastern districts of Sangalla, Makale and Mengkendek; and lower nobles, called *tana' bassi* ('iron stake') or *to makaka* ('freemen' – Waterson 2009, 159–172). Only noble families had *tongkonan* houses and *liang pa'* tombs. Social hierarchy, including within noble classes, is key to understanding ritual life, especially funerals, whose degree of elaboration must reflect the specific rank and status of the deceased, following a very specific ranked scale of funeral categories (see below). Slavery was abolished with Dutch colonisation and social boundaries between nobles and commoners are more blurred today than in the past, but funerals still play an important role in reproducing social status and organisation.

Figure 1.9: Front of a tongkonan *(kinship house) at Karuaya, with stacks of buffalo horns resulting from sacrifices at past funeral ceremonies (photo: G. Robin).*

The social organisation of the Toraja corresponds to the 'house society' kinship model as originally defined by Levi-Strauss (1975). In this model, individuals are not organised in descent groups or lineages but are members of 'houses'. The house kinship group is materialised by a physical origin house, the *tongkonan*, which, unlike more common houses, has a distinctive curved roof and is highly decorated with wood-carved motifs and displays of horns from sacrificed buffaloes (Fig. 1.9). The house has the status of a moral person, and owns lands, heirlooms and, in the past, slaves. Members of the house do not own this wealth but have access to it (Joyce and Gillespie 2000).

An important element in Toraja house society is that each 'house' grouping has both a physical origin house (*tongkonan*) and a rock-cut tomb (*liang pa'*) in which members of the house have a right of burial. The house and the tomb are considered to form a 'pair' (*sipasang*), and no *tongkonan* is really complete without its *liang* (Waterson 1995, 207). The strong ties between houses and tombs are also expressed in ritual poetry, in which *liang pa'* are referred to as 'houses without smoke' (*tongkonan tangmerambu*). The wooden doors of the tombs are decorated with motifs found on houses, such as old *pa'sussu'* vertical grooves (Nugraha 2019) and *pa'tedong* buffalo heads (Waterson 1988). Each *liang pa'* in Tana Toraja is associated to a specific *tongkonan*

Figure 1.10: Erection of a standing stone at a rante *(ritual plaza) in Parinding (Sesean) as part of a funeral ceremony (photo: G. Robin).*

house, located nearby, and is often named after it (*e.g. liang to Bunna'*, 'the tomb of the people from [*tongkonan*] Bunna"). This principle is key to understanding how tombs are distributed in the landscape and how they are used for burials.

Toraja people were traditionally animist. Although a large part of the population has now converted to Christianism, many ritual traditions based on the traditional animist religion (*aluk to dolo*, 'way of the ancestors') have remained and still mark social life today. The traditional religion is divided into two main parts: rites of the East, associated with life and fertility, and rites of the West, associated with death and the ancestors (Nooy-Palm 1986). Most rituals involve the sacrifice of an animal, from chicken to pigs and buffaloes, representing offerings to the deities of nature (*deata*) or the ancestors (*to dolo* or *nene'*).

Death and the ancestors are central to Toraja social life and beliefs. A dead individual has the potential to become an ancestor who will look after the prosperity and fertility of their descendants, as long as respect is shown to them through regular rituals set by the traditional *aluk to dolo* religion (Waterson 2009). Funeral feasts are part of this practice, at the occasion of which commemorative standing stones (*simbuang batu*) are erected in dedicated ritual plazas (*rante*) located close to settlements (Fig. 1.10).

Toraja funeral ceremonies

Funerals represent a major element of the Toraja tradition, even though the importance of the ancestors has declined under the influence of conversions to Christianism. The ceremony not only enables a deceased person to accomplish a process of transition (from the world of the living to Puya, the world of the dead), it also enables social statuses to be expressed and reinforced, with all members of the kinship group standing to benefit from the wealth and prestige displayed at this occasion. Toraja funeral ceremonies are highly elaborate, and the degree of their elaboration is based on the wealth and nobility rank of the deceased. The scale of elaboration is very precisely codified, with different ranks of funeral ceremonies prescribed by the traditional custom (*aluk to dolo*). The funeral ceremonies of noble individuals have a duration of several days. Ceremonies are carried out over a single stage (for lower rank funerals) or two stages (for higher rank funerals). Each level of funeral is marked by an increasing number of animal sacrifices (*e.g.*, from three to 120 buffaloes), and by the addition of ceremonial apparatuses (*saringan* palanquins, *tau-tau* effigies of the dead, *bala'kayan* meat distribution platform, *lakkean* corpse platform, *etc.*). Toraja funeral ceremonies have been studied in detail by several Western visitors and social anthropologists, in particular by Kruyt (1924), Nobele (1926), Koubi (1982), Nooy-Palm (1986) and Waterson (2009, 373–394, 456–457). Only the main aspects will be summarised here, with a focus on the temporality and the spatiality of these rituals, from the biological death of the person to its burial in the *liang pa'*.

When a Toraja noble dies, the person is not considered dead until the funeral ceremony takes place, which normally happens months or years afterwards. In the meantime, the person is considered 'sick': the body is kept in the back room of the *tongkonan* house, is offered food and is visited by relatives. The latter enter a phase of preparation and negotiation to determine the appropriate rank of funeral ceremony and to assemble the resources (especially animals) required to host it. Funeral ceremonies traditionally could only take place during the months of August and September, which is the period between the harvest and the planting of rice. Nowadays, ceremonies occur year-round. All members of the *tongkonan* group of the deceased have an obligation to attend the ceremony, and audiences can gather hundreds of persons, for whom a temporary village is built inside the hamlet, with bamboo shelters that are assembled in the weeks before the ceremony. Once all the necessary conditions are met, the funeral ceremony commences. Most of the rituals take place inside the hamlet (*tondok*) where the deceased body awaits. The sacrifices of the first buffaloes mark the beginning of the funeral. The subsequent rituals are performed by two main actors. The ritual priest, *to minaa* ('the one who knows'), who is of noble rank, supervises all the steps of the ceremony and performs all the prayers and ritual speeches. The *to mebalun* ('the one who wraps') is responsible for preparing the body (wrapping it in multiple layers of cloths for mumification) and remains silent during the entire ceremony. The *to mebalun* was traditionally a slave who, because of the pollution associated with death, lived on the margins of society and was only requested at the time of funerals (Volkman 1979b).

The first stage of the funeral ceremony involves the sacrifice of buffaloes and pigs at the *tondok*, and the wrapping of the deceased body in the back room of the *tongkonan* house. During the second stage of the ceremony, which also spans several days, ritual apparatuses such as the *tau-tau* (effigy of the dead) and the *saringan* (palanquin to carry the dead during the final procession) are prepared and assembled. The wrapped body is then moved from the back room to the middle room of the *tongkonan* and is placed in a north–south orientation (in the direction of Puya), thus marking its official death (the person is no longer considered as sick). The body is then taken out of the house and is placed in the platform under the rice barn located in front of the *tongkonan*, where the *tau-tau* had been placed. All these steps are marked by various prayers and sacrifices of chicken, dogs, pigs and buffaloes.

The wrapped body is then placed on the *saringan* palanquin (Fig. 1.11) and is carried outside the *tondok* to the *rante* (ceremonial plaza). The procession is led by the *to mebalun* (who carries personal effects of the deceased) and includes relatives and buffaloes. The procession stops at the *rante*, where a *lakkean* platform has been built to expose the wrapped body, under which the *tau-tau*, representing the 'soul made visible' of the dead, is placed to watch the next part of the ceremony. At the *rante*, the largest number of buffaloes are sacrificed, and the meat is formally distributed to members of local *tongkonan* groups according to their rank. The sacrifices mark the departure of the soul of the deceased to Puya, the land of the dead. At high-rank

Figure 1.11: House-like saringan (palanquin) with a wrapped body ready to be carried outside of the tondok of Tandung La'bo (Sanggalangi) as part of a high-ranked funeral ceremony (photo: G. Robin).

funerals, a standing stone (*simbuang batu*) is dragged and erected at the *rante*, joining other standing stones erected at previous funerals (Fig. 1.10).

The final step in the ceremony is the procession from the *rante* to the *liang pa'* cemetery. The procession again is led by the *to mebalun* and normally includes only close relatives of the deceased. In contrast with previous phases of the funeral, the last procession is marked by expressions of joy and happiness, reflecting the satisfaction of having accomplished the prescribed funeral process and the appropriate sacrifices. A last buffalo or pig sacrifice is made before the placement of the body inside the rock-cut tomb, the latter act being carried out without particular ceremony. Once the burial is over, closure rituals can take place, which are aimed at purifying the *tondok* and its inhabitants from the pollution of death, and ending the mourning period of the deceased's relatives, who can reintegrate into their normal lives.

Funeral ceremonies vary across areas of Tana Toraja and according to the rank of the noble deceased. However, they share three key elements: first, funerals are not just a single event, but rather a prolonged process that involves multiple steps, gradually marking the transition of the deceased from the world of the living to the world of the dead; second, funerals are all associated with multiple animal sacrifices and feasting; third, funerals employ multiple locations (*tondok*, *rante*, cemetery).

Many aspects of Toraja funerals and other traditional rites described in the ethnographic literature or mentioned by Toraja people today have been impacted by significant changes in the last century. It is important to stress differences between traditional practices as they were carried out 'in the past' and 'now' (Waterson 1993). Several factors are responsible for changes in both social organisation and ritual customs. The short colonial period by the Dutch (1906–1946) led to a sharp decrease in inter- and intra-community warfare and triggered the end of slavery. The establishment of the State of Indonesia and, in particular, the era of the 'New Order' (1966–1998) prompted an economic boom in Sulawesi as in the rest of the country and triggered a major inflation in ritual expenditures as well as a wider participation in prestigious funerary practices across the Toraja population. Finally, the Christianisation of the Toraja population led to a decline in traditional beliefs. Conflicts emerged between the Toraja church (Geraja Toraja), of protestant obedience, and aspects of funeral practices, such as the making and curation of *tau-tau* effigies, perceived as idolatry. These conflicts are now settled and traditional ceremonies are more tolerated by the church. Funerals are still practised according to traditional prescriptions, although Christian prayers can be added, and most Toraja participants no longer believe in the religious meaning of the rites. While the social element has survived (display of rank and wealth), the religious one no longer prevails (the making of ancestors to whom blessing may be asked). As a result, traditional funerals are still commonly celebrated today, and animal sacrifices are still carried out; however, they are no longer addressed to the ancestors and the deities as they used to be.

Rock-cut tombs have also been impacted by these societal and religious changes and indeed offer interesting insight into these changes. This book mainly focuses on the traditional aspects of *liang pa'* but also addresses the impact of recent changes on the creation and uses of the tombs today.

Studies on Toraja stone monuments: from ethnography to ethnoarchaeology

The Toraja rock-cut tombs and standing stones are part of a wider Austronesian culture of stone monuments, which have developed over the last few centuries and some of which are still living today. These monuments are often classified by archaeologists as 'megalithic': they are made of large stones that are quarried and dragged to a certain place to create commemorative and funerary architectures. Their construction requires a cooperative labour force assembled at particular events such as feasts. Most famous examples are the megalithic chambered tombs of the Merina in Madagascar (Bloch 1971) and of West Sumba, Indonesia (Hoskins 1986), the standing stones of North-East India (von Fürer-Haimendorf 1939) and of several Indonesian islands (see Sherman 1990 for the Batak of Sumatra; Beatty 1992 for Nias; Forth 2001 for Flores).

These ethnographic contexts and monument building traditions were first examined by travellers and social anthropologists. From the 1990s, they have attracted the attention of western archaeologists interested in prehistoric monumentality and associated rituals in European contexts. Studies on 'ethnographic megaliths' have sometimes been used as comparative models to reinterpret European prehistoric monuments (*e.g.*, Gallay 2006; Jeunesse *et al.* 2016). Such ethnoarchaeological approaches have been done either indirectly (using anthropological literature) or directly, through enquiry in the field (*e.g.*, Parker Pearson *et al.* 2010; Adams and Kusumawati 2011; Hayden 2016; Steimer-Herbet 2018; Adams and Robin 2022; Jeunesse *et al.* 2022).

Toraja *liang pa'* are not considered megalithic monuments, and perhaps for this reason have received less attention than their megalithic counterparts across the Austronesian world. The relative lack of research on *liang pa'* is certainly not due to a lack of interest in the Toraja culture itself. Toraja culture and society have been extensively studied ethnographically since the 1960s. Anthropological studies have focused on Toraja social organisation (Waterson 1986; Bigalke 2005; Buijs 2006), worldviews and death rituals (Nooy-Palm 1979; 1986; Koubi 1982; Volkman 1985; Jannel and Lontcho 1992; Waterson 1993; Sandarupa 1997; Tsintjilonis 2000; 2007), oral traditions (Waterson 1997; Sandarupa 2016) or, more recently, on the impact of tourism (Adams 2006) or the financial crisis (de Jong 2013). *Liang pa'* rock-cut tombs are frequently mentioned in that literature (most detailed accounts are: Grubauer 1913, 200–203, 214–220, 245–246, 258–259; Kruyt 1924, 161–169; Keers 1939; Wilcox 1949, 84–85; Nooy-Palm 1979, 259–261; Brisbois and Douvier 1980, 116–117; Koubi 1982, 193–196, 229–230; Crystal 1985, 141–143; Waterson 1990, 199–228). However, they have never been the primary focus of anthropological research. A few ethnoarchaeological studies have been initiated in the last 20 years in Tana Toraja (*e.g.*, Adams 2004). Recent fieldwork research by Akin Duli (Duli 2015; 2018; Duli *et al.* 2019) and Christian Jeunesse (Jeunesse and Denaire 2018) have highlighted traditional burial sites and have been particularly worthwhile at addressing this knowledge gap. However, specific questions have not been addressed in the literature and a more systematic approach to rock-cut tombs, including surveys of multiple areas (rather than a few selected case studies), has been needed.

The main gaps in the literature concerns the following primary issues and questions:

1. *The social meaning of rock-cut tombs*: why do Toraja people use rock-cut tombs, instead of other forms of burials (*e.g.*, megalithic tombs, as most other Austronesian societies)? Why do Toraja value rock-cut tombs? What are the other forms of burials in Tana Toraja, and how are they perceived in comparison to *liang pa'*?

2. *The architecture and decoration of the* liang pa': how large are the chambers? Do they vary across Tana Toraja localities? Can tombs have multiple chambers? What are the main decorative motifs found on the door of the tombs? What is their meaning? Are there other forms of decoration, for instance on the walls inside the chambers? Have *liang pa'* architecture and decoration changed over time since the 17th century?
3. *The construction of the* liang pa': on what occasions are new tombs constructed? Who carve them out? What tools and processes are used? How long does it take and how much does it cost to have a tomb cut and decorated? Are there any rituals involved during the process, as there are for the construction of *tongkonan* houses?
4. *Burials inside the tombs:* how many persons can be buried inside a tomb? How are bodies placed? Is the position of the bodies inside the tombs determined by social criteria such as gender, age or rank? Are objects deposited together with the bodies? How are the multiple depositions managed over time? Can older depositions be displaced or taken out of the tomb? What happens when a tomb is full? Are there ritual depositions in front of or inside the tombs in between funerals?
5. *The landscape setting of* liang pa': where are tombs located in the landscape? How far are they from the hamlets? Can tombs be built anywhere in the landscape or are there any customary prescriptions about their visibility and distance from houses, cultivated lands, roads? How does variation in geology and landform influence their landscape setting? Are there any patterns in the landscape settings of the tombs which can inform us on the nature of the relationship between Toraja people and their ancestors?
6. *Cemeteries and kinship systems*: can a kinship group (*tongkonan* house) have more than one tomb? Can they have tombs in several cemeteries? Do cemeteries belong to single hamlets or are they shared by several hamlets? Is there any competition between kinship groups regarding the location of their tombs inside a cemetery? As Toraja people normally have memberships in more than one *tongkonan* group, in which *tongkonan*'s tomb are they eventually buried?
7. *The impact of recent changes on the* liang pa' *tradition*: how have the significant economic, religious and social changes in Tana Toraja since the 1970s affected practices related to the architecture, decoration, landscape setting and kinship uses of the tombs?

Fieldwork 2017

To address these gaps, the two authors of this book carried out a three-week fieldwork study in Tana Toraja in June 2017. Our methodology combined site surveys and interviews with local informants, reflecting traditional ethnoarchaeological practices (David and Kramer 2001). Our main objective was to visit as many traditional villages and burial sites as possible to collect a maximum of data. We worked with Toraja guides who facilitated access to areas and communication with local communities.

Our fieldwork focused on three main areas from within a 20 km radius around our base in the town of Rantepao (Fig. 1.5). The first area (A) is located to the north of Rantepao and covers the districts on the lower slopes of Mount Sesean (Sesean, Sesean Suloara', Tikala Suloara' and Tallunglipu). The area is marked by volcanic boulders scattered over the landscape and inside which *liang pa'* are cut. The second area (B) covers districts to the south of Rantepao, on the eastern bank of the river Sa'dan (Kesu, Makale Utara, Sangalla, Sangalla Utara). The area is dominated by sedimentary reliefs of the Makale formation, with high limestone cliffs suitable for the establishment of *liang pa'* cemeteries. The third area (C) is located to the west of Makale and the river Sa'dan and covers the northern part of the district of Rembon, as well as parts of the districts of Malimbong Balepe' and Rantetayo. The area has volcanic outcrops and boulders.

In total, we visited cemetery sites in 21 *lembang* or *kelurahan* territories (administrative units that compose the districts, and which correspond to a group of hamlets). Combined, the cemetery sites represent a total of *c.* 650 rock-cut tombs. We mainly used GPS, photographs and sometimes photogrammetry to record the context of the tombs, and the tombs themselves (in particular their decoration and, where possible, their internal spaces). The majority of the cemetery sites we visited included older and more recent rock-cut tombs, which enabled us to observe patterns of change through time. Other tombs were in the process of being cut during our visit, which provided information on the cutting work of traditional *liang pa'*. We also carried out semi-structured interviews with stone workers at tomb cutting sites in cemeteries, tomb owners and sponsors, and ritual specialists (*to minaa*). Where possible, we visited *tondok* hamlets and recorded information on the *tongkonan* groups associated with specific tombs.

Our fieldwork was, by nature, partial and fragmentary, but enabled us to collect new information which, combined with the existing literature, made it possible to address the seven key issues outlined above. The main aim of this book is to provide a synthesis of knowledge on the *liang pa'* monuments by presenting their primary characteristics and by discussing their use and social significance. Such a task inevitably presents several challenges. The first is the geographical diversity of the material and changes that have affected it through time. Rock-cut tombs are made and used in slightly different ways throughout Tana Toraja and these practices have changed over time. Often in this book, we indicate the specific area of such and such traits, and whether they are more relevant to traditional customs or to more recent practices. The second challenge is the fragmentary nature of our information. This book relies on literature sources and on limited direct field observations. It cannot claim to provide an exhaustive description of the rock-cut tombs and associated practices in Tana Toraja. Instead, we hope it provides a clear and synthetic overview of the *liang pa'* tradition as a whole, while providing all the necessary details and literature references, and raising broader points on the use of rock-cut architecture as a social strategy of commemoration and identity.

Acknowledgments

We would like to thank Roxana Waterson and Dimitri Tsintjilonis for very useful discussions and help with preparing the fieldwork, our Toraja guides Amos Palungan and Samuel Palangda', and the people who kindly provided useful information on *liang pa'* matters in Tana Toraja: Abraham Sulu', Daniel Sulle, Daud Moning Tulak, Julius Kamma, Markus Tandikaloden, Martin Bakkan, Ne' Dane', Nene' Limbong, Ne' Sampe, Pak Saipan, Yus Paulus Senga, Paulus Kondosara, and Rombe'. Fieldwork in Tana Toraja was funded by the Challenge Investment Fund, University of Edinburgh.

Plate 1: Cliff cemetery at Lemo (Makale Utara), with rectangular entrances to rock-cut tombs and tau tau *human effigies of the dead in rock-cut galleries (photo: G. Robin).*

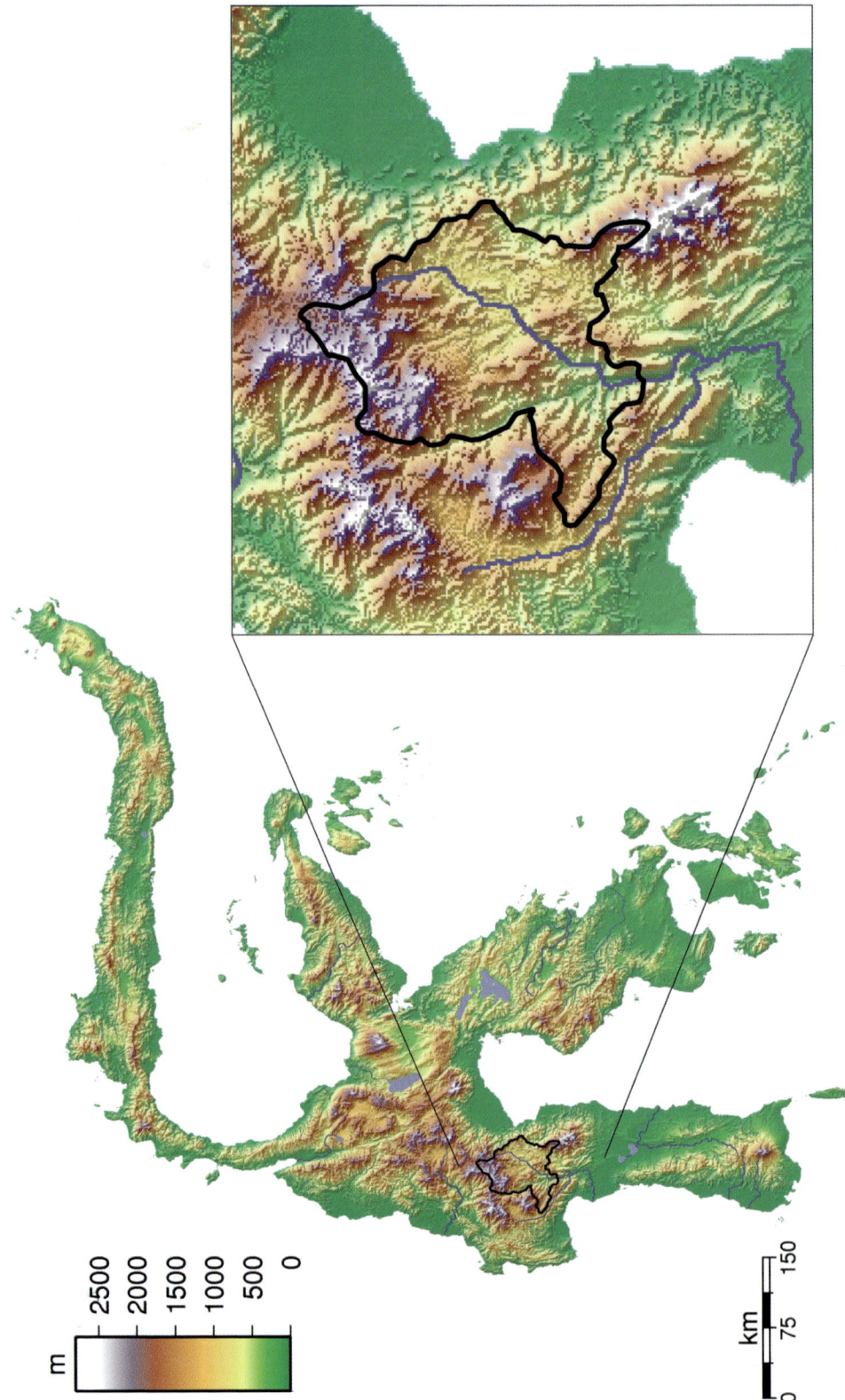

Plate 2: Topographic map of Sulawesi, with a close-up view of the Tana Toraja area (map: G. Robin; basemap: Wikimedia CC BY-SA 3.0).

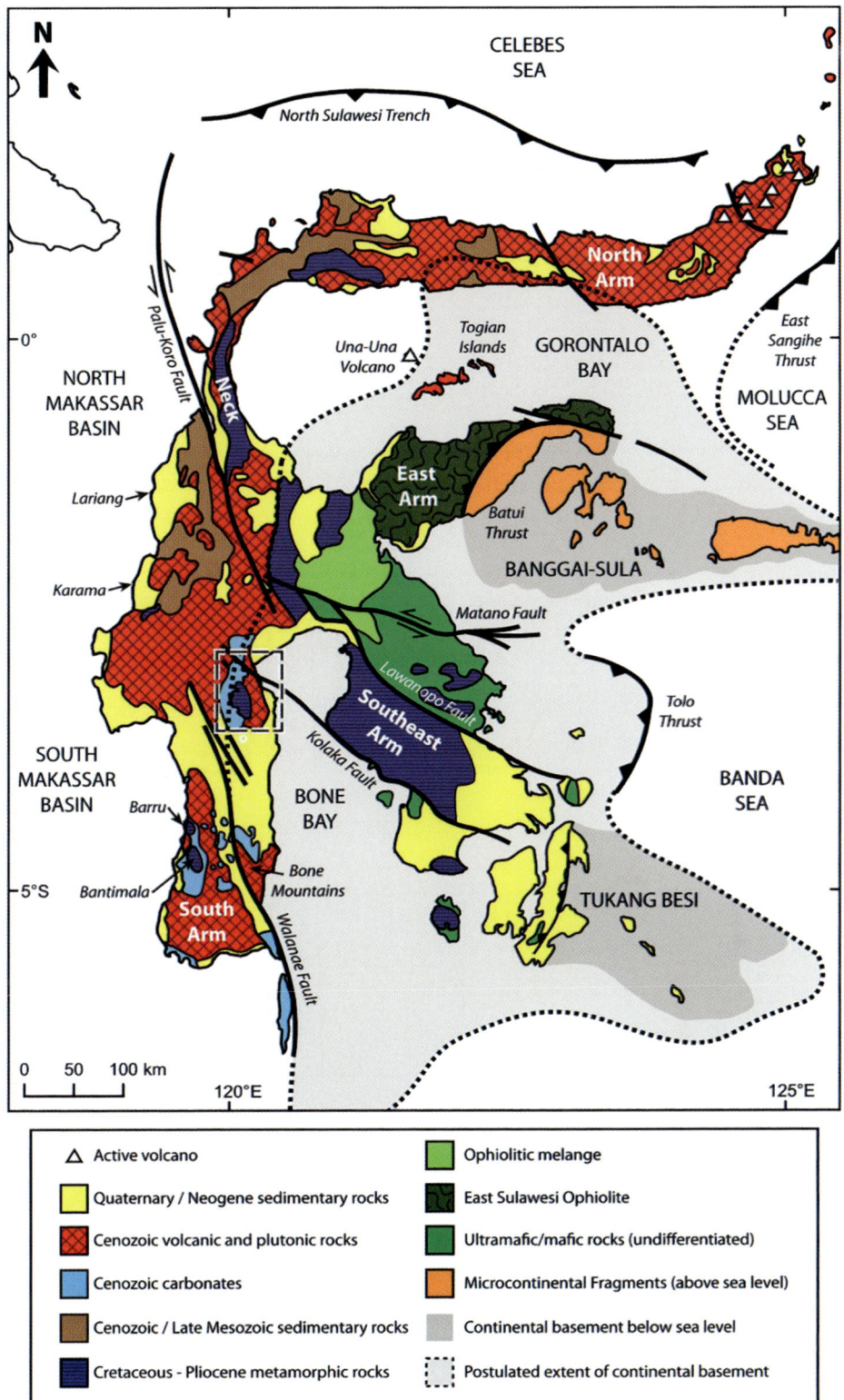

Plate 3: Geological map of Sulawesi (after White et al. 2017).

Plate 4: Volcanic landscape on the slope of Mount Sesean, with basalt outcrops and boulders. Note the rock-cut tomb in the process of being cut at the centre of the image (Lempo, Sesean Suloara'; photo: G. Robin).

Plate 5: Main decorative motifs used on the wooden doors of liang pa' *rock-cut tombs.*

Plate 6: Recent liang pa' *rock-cut tombs with decorated wooden doors on Bori' boulder 21 (Sesean). The two lower tombs have a pa'tedong (buffalo head) motif, while the two upper ones have a pa'barre allo (sun) motif (photo: G. Robin).*

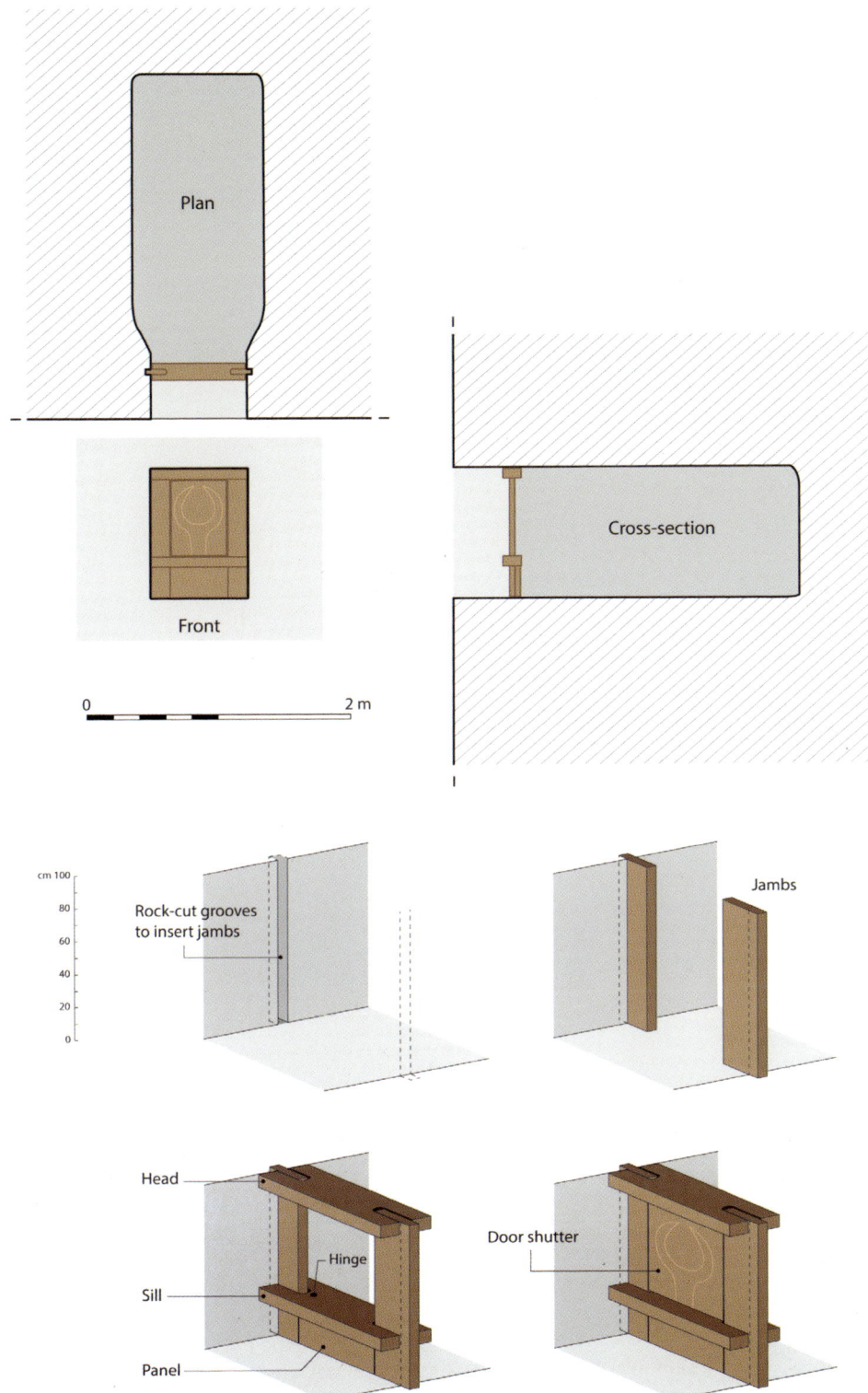

Plate 7: Anatomy of older liang pa' *rock-cut tombs (style 1). Top: plan, cross-section and front elevation; bottom: construction of the wooden door (diagram: G. Robin).*

Plate 8: Entombment at Lempo boulder 18 (Sesean Suloara'): the deceased's body, wrapped in a cylinder of cloths, is carried down the rock to the entrance of the rock-cut tomb (photo: G. Robin).

Plate 9: Location of To Bua' cemetery (Deri) in relation to local tongkonan houses (white rectangles) and other boulder cemeteries (yellow dots) (map: G. Robin; satellite image: Google).

Plate 10: Satellite view of Ke'te' Kesu (Kesu), with locations of its tondok (village), rante (ceremonial plaza) and cliff cemetery (map: G. Robin; satellite image: Google).

Plate 11: Satellite view of Buntu Pune (Kesu), with locations of its cliff cemetery and ancient and current tondok (village). The white arrow represents the path from the current tondok to the cemetery (map: G. Robin; satellite image: Google).

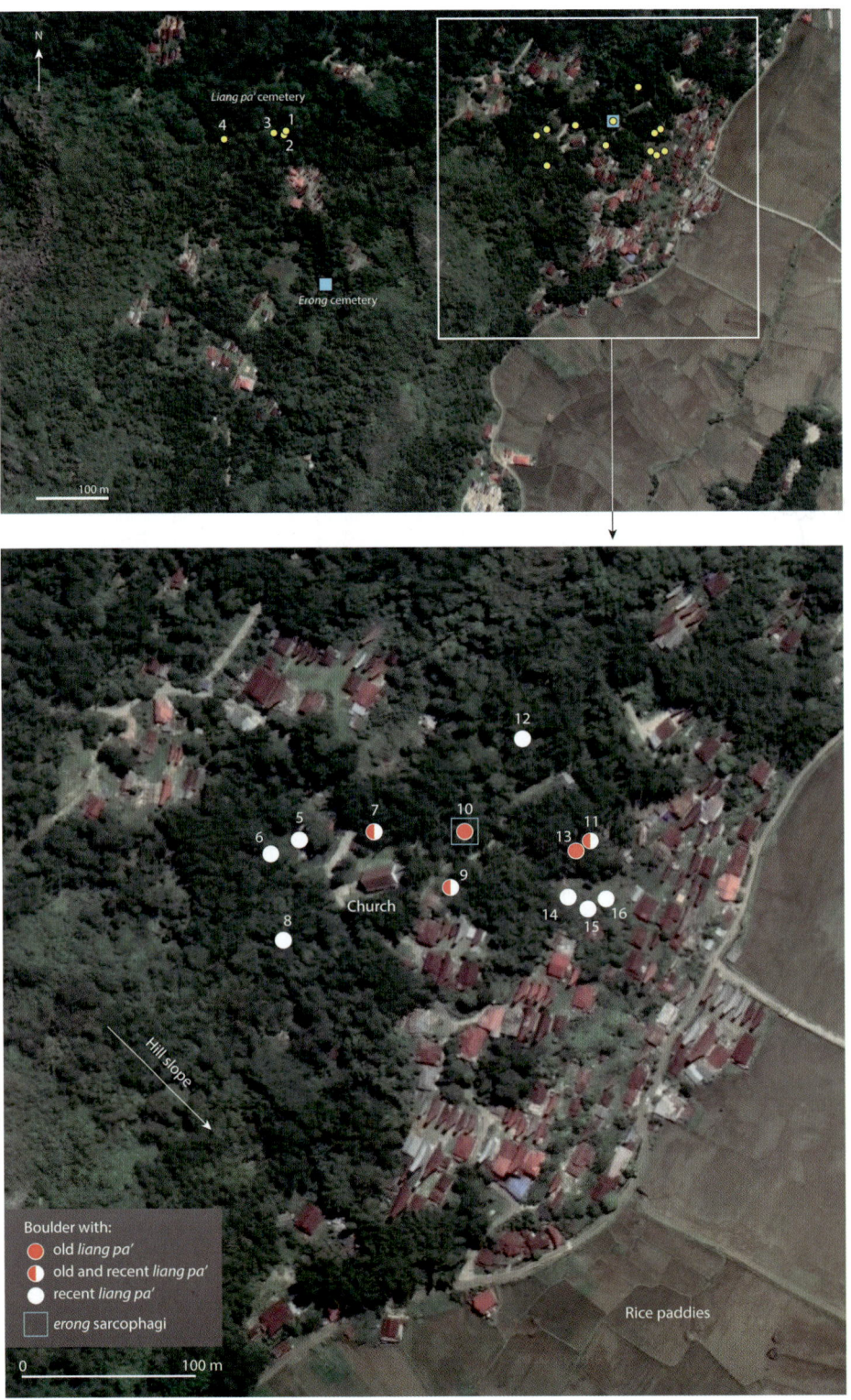

Plate 12: Distribution of boulders with liang pa' *rock-cut tombs and* erong *sarcophagi in Parinding (Sesean) (map: G. Robin; satellite image: Google).*

Plate 13: Distribution of boulders with liang pa' rock-cut tombs in Bori' (Sesean) (map: G. Robin; satellite image: Google).

Plate 14: Distribution of boulders with liang pa' rock-cut tombs in Buntu Lobo (Sesean) (map: G. Robin; satellite image: Google).

Plate 15: Rock-cut tomb cemetery of Batu Lappa' (Buri', Rembon) (photo: G. Robin).

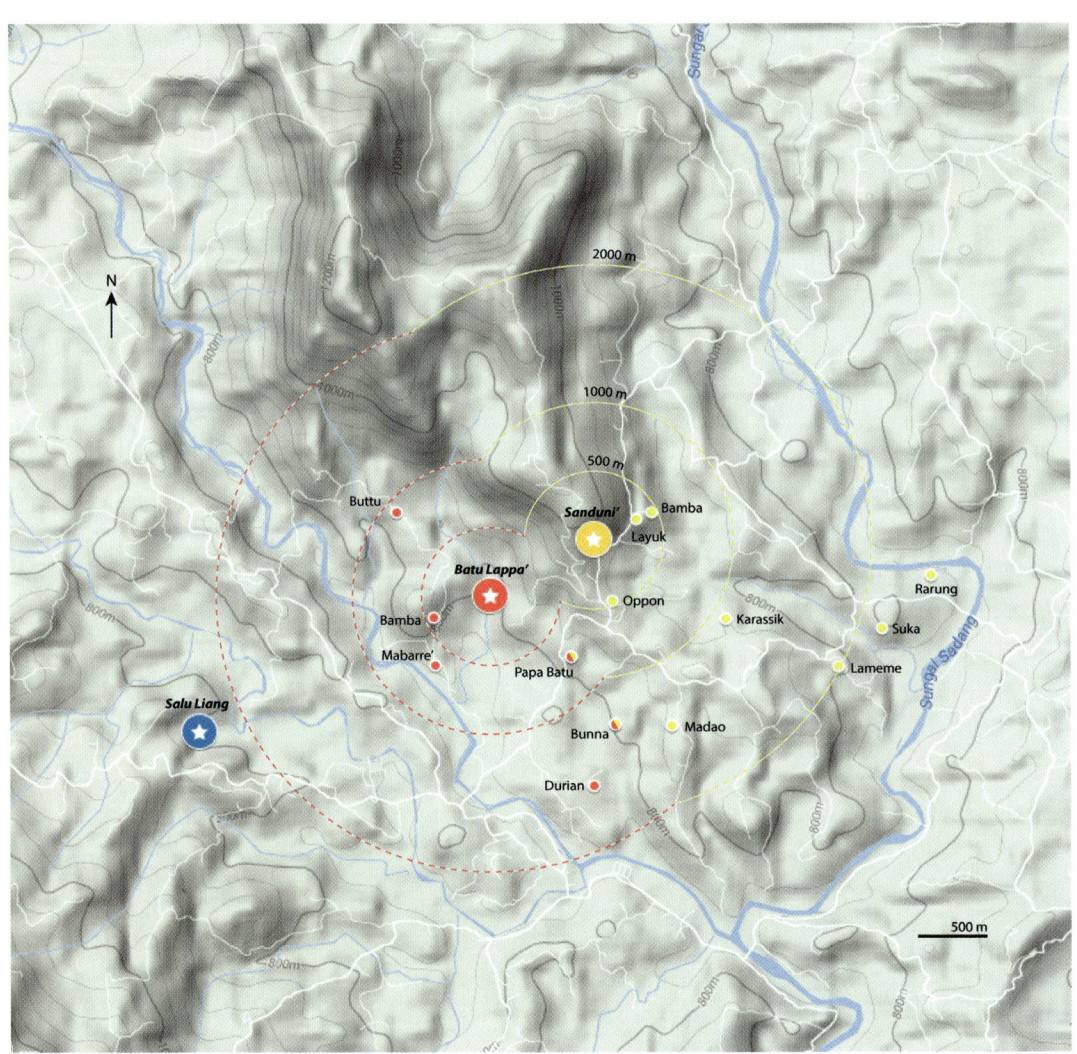

Plate 16: Location of tongkonan *kinship houses associated with the rock-cut tomb cemeteries of Batu Lappa' and Sanduni' (Rembon) (map: G. Robin; basemap: Google).*

Plate 17: Rock-cut tomb cemetery of Sele in Suolara' (Sesean Suloara'). Top: elevation (orthophoto from photogrammetry survey); middle: distribution of used and unused tombs (as per June 2017); bottom: date ranges for the creation of the tombs (images: G. Robin).

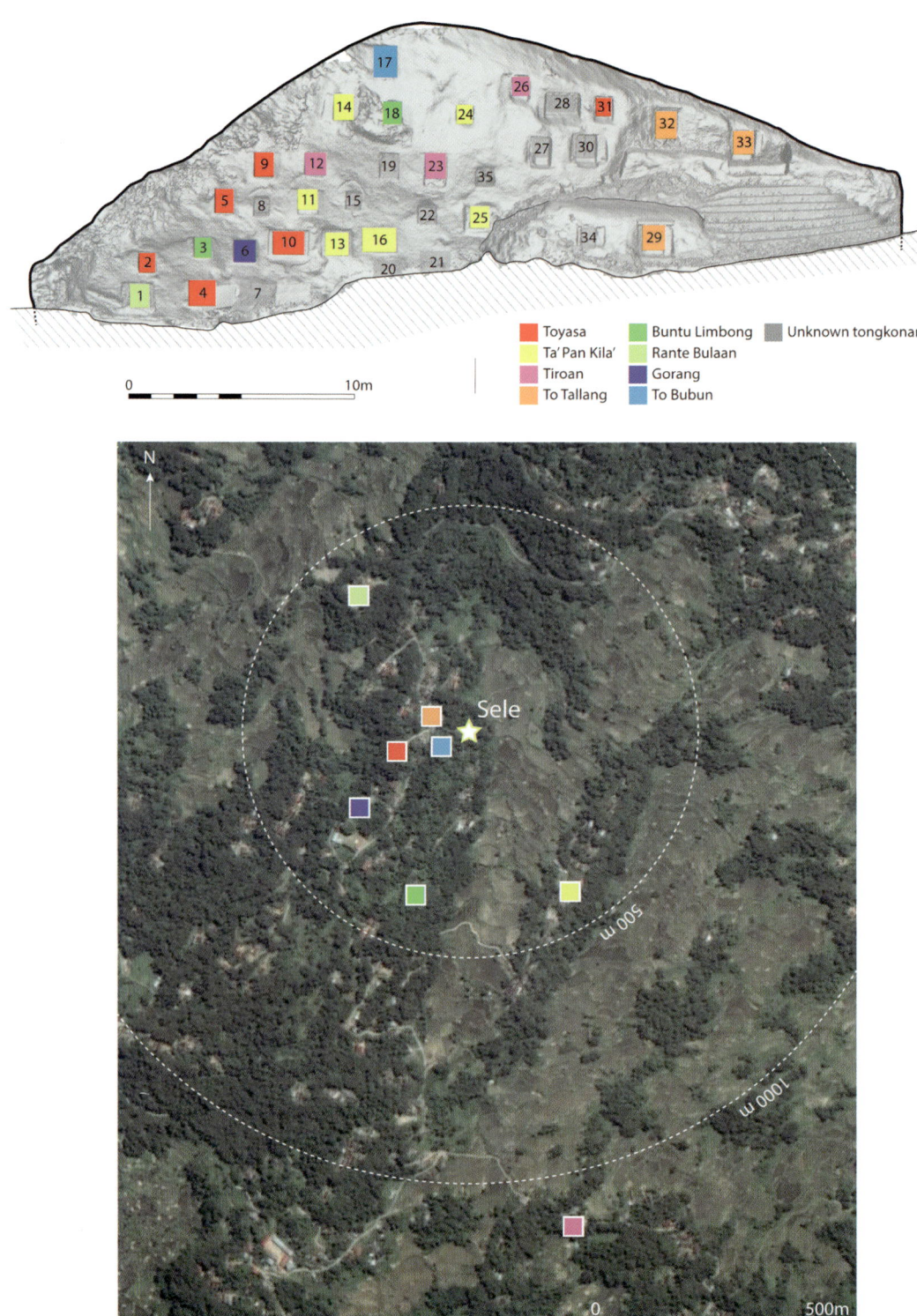

Plate 18: Tongkonan kinship houses associated with the cemetery of Sele. Top: location of tombs with their corresponding tongkonan house affiliation; bottom: location map of tongkonan houses (images: G. Robin).

Chapter 2

Tomb traditions through time and space in Tana Toraja

The dead in Tana Toraja are considered more important than the living. The ancestors are believed to retain agency over the living and to be a source of fertility (Waterson 2009). It is therefore essential for kinship groups of higher ranks to preserve the physical remains of the dead and to keep them in special places and formal tombs. The *liang pa'* rock-cut tombs represent a common type of aristocratic tomb but other burial traditions are known in Tana Toraja. Some of them belong to the past and are no longer used, while others have developed recently or are unique to certain areas within Tana Toraja. The aim of this chapter is to give an overview of these various resting places for the dead, which form the broader funerary context within which the rock-cut tombs hold a particular place.

There are three main types of aristocratic burials in Tana Toraja: wooden sarcophagi, rock-cut tombs and house-tombs (Keers 1939; Koubi 1982, 195–196; Duli *et al.* 2019). They all share the trait of being collective burial monuments, each associated with a specific *tongkonan* kinship house. Other forms of burial will also be discussed at the end of the chapter. In addition, burials for individuals of lower rank, which consist of simple, in-ground burials or depositions in caves, will be presented though not discussed extensively.

Wooden sarcophagi (*erong*)

The oldest known aristocratic burial tradition in Tana Toraja, which is extinct today, entailed grouping the dead into ornamented wooden sarcophagi[1] (Fig. 2.1). These were displayed outside of villages in specific landscape contexts. In the western region of Mamasa, they are found on hilltop locations and were traditionally protected under an open, wooden shelter (Duli 2014). In the Sa'dan region, they are called *erong*[2] and are found in high cliffs or in rock shelters. They were normally placed in high locations, either on natural anfractuosities in the cliff (*e.g.*, horizontal crevices), suspended with ropes, or placed on wooden supports inserted into the wall of the cliff. All these techniques can be combined at single sites. Such high positions are used for these burials in order to protect them from theft. In addition, their conspicuous placement in high elevations can also be seen as a strategy of social display by aristocratic families (Nooy-Palm 1979, 259). The sarcophagi are always grouped together in these specific natural places, which form ancient communal cemeteries. According to recent surveys, 19 such cemeteries are known in the Sa'dan region and 21 in the Mamasa region (Duli 2014; 2015). During our visit in 2017, we observed 11 of them (Fig. 2.2).

The sarcophagi are normally made of a hard wood known as *uru* in the Toraja language (*Magnolia vrieseana*). Their construction consists of two primary elements: a box-like receptacle, and a lid. Some sarcophagi, especially in the region of Mamasa, are made from a single tree log, which is split into two parts: one used for the lid, while the other is used for the receptacle and is hollowed out (Fig. 2.1). The resulting receptacle is very similar to a domestic object called *issong*, which is a canoe-shaped wooden mortar used in villages for pounding rice (Keers 1939, 209; Koubi 1982, 101). In the region of Sa'dan, the receptacles of house-shaped *erong* are occasionally made of five wood planks assembled together. In both regions the sarcophagi are shaped in three main specific forms: house (or boat[3]), buffalo and pig[4] (Fig. 2.6). These distinctive forms are believed to reflect rank or age/gender divisions. Eric Crystal (1985, 141–142), for instance, reports that boat-shaped sarcophagi were used by princely nobles (*Puang*) exclusively, while buffaloes were for middle-rank landowners (*to makaka*) and pigs for slaves serving the nobles (*kaunan*). A similar explanation was given by our informants in Rembon (southwest of the Sa'dan region), except that pig-shaped sarcophagi were designated for commoners (*to buda*) rather than slaves. There seems to be regional variation concerning these rank attributions, as was confirmed by informants from Sesean Suloara' (north of the Sa'dan region): according to these informants, house-shaped *erong* were for the highest nobles, while buffalo shapes were for lower noble males, and pigs for lower noble females. Finally, W. Keers (1939, 211) reports that pig-shaped sarcophagi were used for the children of nobles. In the 11 *erong* cemeteries we visited in June 2017, we found that the house-shaped style is represented in the majority of the sarcophagi, followed by the buffalo style. Pig-shaped *erong* appear to be relatively rare based on our observations.

The surfaces of the house-shaped sarcophagi are often covered in intricate incised decoration, especially in the Sa'dan region (Nooy-Palm 1979, 239–240; Waterson 1988, 37–38). This style of ornamentation is called *pa'erong* ('*erong* carvings' – Koubi 1982, 101): it is specific to sarcophagi and is not found on houses or rice barns. Motifs include spirals, stars, snake-like motifs (representing the *naga*, the underworld serpent), squares and lozenges (Figs 2.7, 2.8, 2.9). A few *erong* display stylised buffalo heads (*pa'tedong*) (Fig. 2.10), a type of motif that is also found on houses and rock-cut tombs (see Chapter 3). At Londa, Buntu Pune, Tampangallo and Lombok Parinding we also observed *erong* decorated with series of incised parallel lines. This particular type of ornamentation, called *pa'sussu'*, is also visible on very old houses and rice

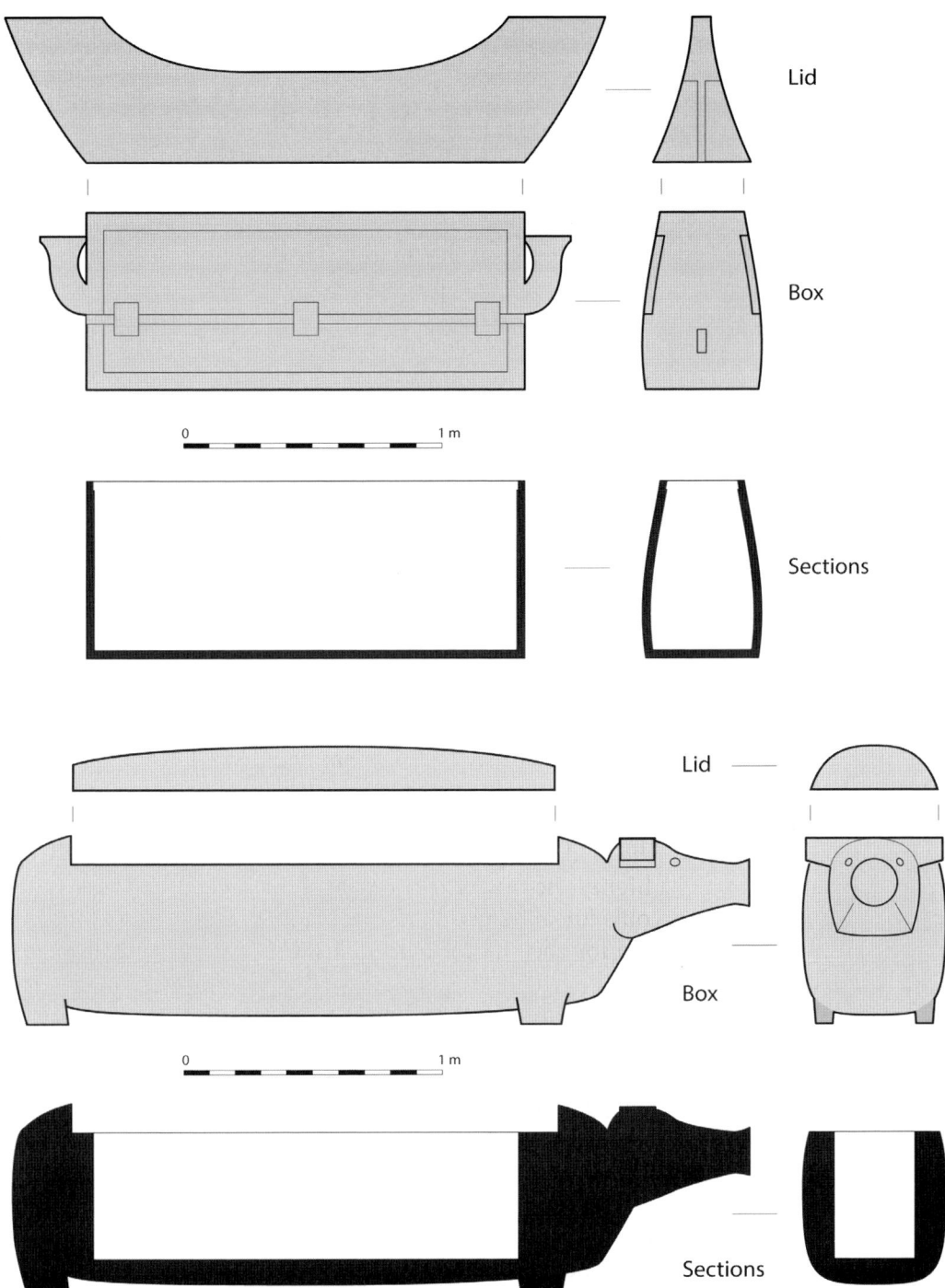

Figure 2.1: Two main styles of erong *(wooden sarcophagi) from the Sa'dan region of Tana Toraja: house-shaped (top) and animal-shaped (bottom) (drawing: G. Robin).*

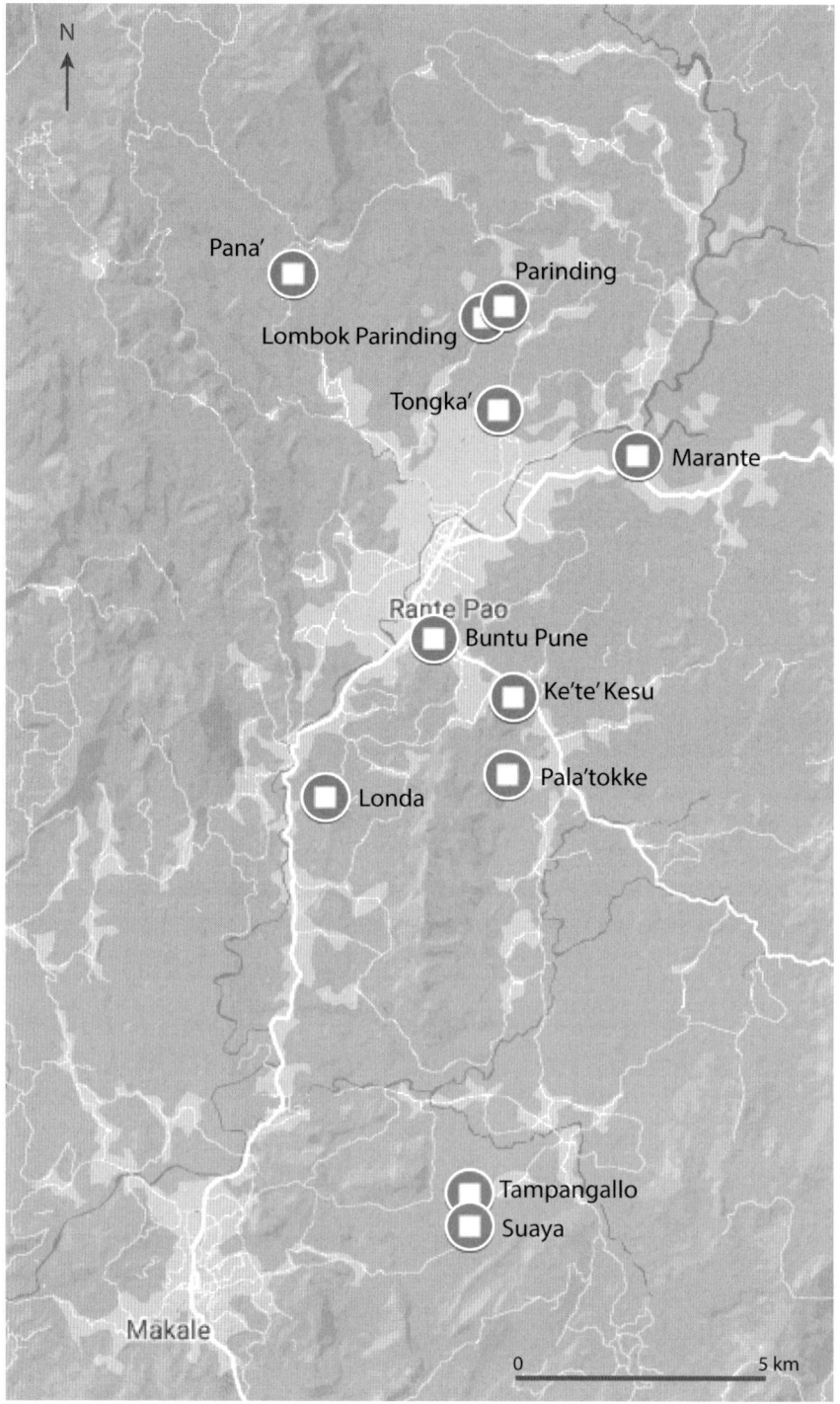

Figure 2.2: Location of erong cemeteries in the Sa'dan region of Tana Toraja (map: G. Robin; basemap: Google).

Figure 2.3: Erong sarcophagi placed on wooden support inserted into the limestone cliff of Pala'tokke (Sanggalangi) (photo: G. Robin).

Figure 2.4: Rock-shelter of Marante (Tondon), which was used as an erong *cemetery (photo: G. Robin).*

Figure 2.5: Rock-shelter erong *cemetery of Lombok Parinding (Sesean), with stacks of old abandoned sarcophagi (photo: G. Robin).*

barns (Fig. 2.11). Each sarcophagus presents a different combination of motifs, which probably helped families recognise their *erong* among the many others in the cemetery.

Erong cemeteries are located within a range of 100–150 m from the nearest associated hamlets. These rock-shelters or rock faces are often located in discreet locations in the landscape, hidden from distant views by dense forests, and away from main paths or roads. They are nevertheless easily accessible, and some of them even become signposted touristic attractions (*e.g.*, Londa, Tampangallo, Pana', Ke'te' Kesu, Lombok Parinding). Cemeteries were used more-or-less intensively by ancient local noble populations. For instance, the *erong* cemetery we saw in Parinding presented only four sarcophagi, while at Lombok Parinding, dozens were piled up at the bottom of the rock shelter. The *erong* we observed at different sites presented various states of preservation: some were badly fragmented, while others were remarkably well preserved. Such variation reflects the very long duration of this funerary tradition in Tana Toraja. Dr Akin Duli (University of Makassar) has recently carried out a radiocarbon dating programme on wooden sarcophagi from both the Mamasa (Duli 2014) and Sa'dan regions (Duli 2015). The results show that they were in use as early as the 9th century AD, while the most recent samples were dated to the 1960s–1970s. Wooden sarcophagi have been proven to represent a 1000-year tradition.

Figure 2.6: Top: house-shaped erong *at Tampangallo (Sangalla); bottom: buffalo-shaped* erong *at Lombok Parinding (Sesean) (photos: G. Robin).*

Figure 2.7: Incised ornamentation on the side of an erong sarcophagus at Tampangallo (Sangalla) (photo: G. Robin).

Figure 2.8: Erong *sarcophagi with incised ornamentation at Lombok Parinding (Sesean) and Ke'te' Kesu (Kesu) (photos: G. Robin).*

2. Tomb traditions through time and space in Tana Toraja

Figure 2.9: Erong sarcophagi with incised ornamentation at Tampangallo (Sangalla) and Ke'te' Kesu (Kesu) (photos: G. Robin).

Figure 2.10: Pa'tedong *(buffalo head) motif incised among geometric motifs on an* erong *sarcophagus at Lombok Parinding (Sesean) (photo: G. Robin).*

Figure 2.11: Top: erong *sarcophagus with* pa'sussu' *(vertical grooves) ornamentation at Tampangallo (Sangalla); bottom:* pa'sussu' *decoration on a rice barn at Karuaya (Sangalla Utara) (photos: G. Robin).*

Such surprisingly old dates certainly explain the very disturbed nature of the sarcophagi's burial contents. *Erong* sites today are often marked by hundreds of human bones dispersed on the ground, some of them re-assembled (*e.g.*, lines of skulls) and tidied into or around the wooden receptacles. Many sarcophagi were originally suspended on cliffs and eventually fell to the ground, scattering the human bones and mixing them with those from other fragmented *erong*. Local populations progressively lost track of which bones belonged to which sarcophagi,[5] resulting in the rather random skeletal assemblages that we can see today. Due to this taphonomic process, it is difficult to determine how these burial receptacles were used in the past. Duli *et al.* (2019, 5–6) claim they were used as ossuaries, *i.e.*, as secondary collective burials, although they do not provide evidence or cite a primary source to support this statement. According to them, the primary burial that preceded the final deposition entailed two phases: first, the body was buried into the ground of the *rante* (ceremonial plaza), before being exhumed and buried again in another location, which was marked by a small cairn (*karopi*). After some time, the bones were collected and placed inside the sarcophagus.

Measurements of *erong* may support the assertion that they were used as ossuaries. According to Duli *et al.* (2019), the average exterior dimensions of an *erong* are 2.0 m in length, 1.0 m width, and 1.20 m in height, while the interior of the receptacle typically measures 1.60 × 0.65 × 0.85 m. Our own measures indicate that *erong* were likely not large enough to accommodate primary inhumations. Measurements from *erong* we observed in Tampangallo had exterior lengths that ranged from approximately 1.50–1.85 m, exterior widths that ranged from *c.* 0.27 to 0.50 m, and exterior heights that ranged from *c.* 0.58 to 0.80 m (Fig. 2.1). The thicknesses of the *erong* walls ranged from 2–5 cm, constraining the interior dimensions of the sarcophagi. The width at the tops of these *erong* were particularly narrow. In one case, the top of an *erong* had an opening that was merely 20 cm wide when accounting for the width of the wooden sides of the sarcophagus. Such a small opening would have been impractical for an adult primary burial and indicates that *erong* were used as ossuaries for secondary burials only.

Despite the uncertainties regarding burial processes, there is a consensus concerning the notion that *erong* were made to receive several bodies over time (Kruyt 1924, 166–167; Koubi 1982, 101; Crystal 1985, 141–142; Waterson 1988, 37) and they were therefore collective burials, probably used by families over generations, which is a typical trait of aristocratic burials in Tana Toraja.

Erong sarcophagi are no longer created and used today as a type of burial receptacle. The sites themselves, however, may still be used for other forms of burial depositions in some areas. At Londa and Ke'te' Kesu, for instance, individuals of high aristocratic ranks can occasionally be buried in coffins with distinctive oval shapes and decorations, which are inserted into small crevices high up in cliff faces (Fig. 2.12). Commoners, on the other hand, are buried in regular (rectangular), undecorated coffins, which are deposited inside karst caves located at the bottom of those cliff faces.

Figure 2.12: Coffin of a recent aristocratic burial placed at a very high cliff crevice at Londa (Kesu) (photos: G. Robin).

Rock-cut tombs (*liang pa'*)

The forms and uses of *liang pa'* will be described in the following chapters. Here we would like to give an appreciation of how this tradition developed through time and how it relates to the other traditions described in this chapter.

Rock-cut tombs are exclusively found in the Sa'dan region of Tana Toraja and seem totally absent in the Mamasa region. These small hewn-out chambers are used as family graves. They are grouped together in limestone cliffs or basalt boulders which form vertical cemeteries. A wooden door is placed to close the open end of each tomb when they are used for burials (Fig. 2.13).

The origin of rock-cut tombs is not entirely clear. When did Toraja nobles start creating rock graves, and why? According to various sources, the tradition emerged in the late 17th century AD and was prompted by two coinciding factors. The first was the repeated incursions by Buginese forces in Tana Toraja, who ransacked villages and *erong* cemeteries (Kruyt 1924, 167). Rock-cut tombs, as discreet, closed, rock chambers, located at inaccessible parts of steep rock cliffs or boulders, were therefore invented as a way to better protect the dead and to secure the valuables placed with them.

Figure 2.13: Rock-cut tomb cemetery of Lo'ko' Mata in Tonga Riu (Sesean Matallo) in 1979 (top) and in 2017 (bottom). Note most recent tombs with distinctive rock-cut balconies and buffalo head, contrasting with the simplicity of older tomb entrances (photos: J. Koubi, with kind permission of CNRS Images; G. Robin).

The second factor was the adoption of metal technologies in Tana Toraja during the same period, which enabled the production of the iron tools required to hew out the solid rock (Nooy-Palm 1979, 259; Crystal 1985, 143; Waterson 1988, 37). Brisbois and Douvier (1980, 116) also propose that the rock chambers, at that time, were adopted as more enduring receptacles than the *erong* sarcophagi, whose wood components naturally deteriorated over time.

The period of the Bugis invasions in the late 17th century is still recounted in the oral tradition today. The Toraja community leaders who united and fought against the invaders are remembered as the 'Ancestors of the Same Dream' (Waterson 2009, 41–60). According to one of our informants, it is one of these ancestors, Songgi Pataki,[6] who encouraged the creation of the first rock-cut tombs, which were made in Lemo in 1680. However, Songgi Pataki himself, we were told, was buried in an *erong* sarcophagus.

While the *liang pa'* rock-cut tombs thus probably emerged around the 17th century, they did not replace *erong* burials (*contra* Crystal 1985, 143). Both traditions co-existed for three centuries (Fig. 2.14), with different dynamics: while the *erong* tradition seems to have declined before being abandoned definitively a few generations ago (Duli 2015), the *liang pa'* have spread throughout the Sa'dan region and have been subjected to stylistic transformations (see below). They are still very dynamic today, with several new tombs cut every year (see Chapter 4).

During our survey in 2017, we identified two main styles of *liang pa'*, which seem to correspond to two main phases of development (Fig. 2.15). The older *liang pa'* tombs (style 1) are relatively small, with an entrance *c.* 0.75 × 0.50 m, and a chamber 1.80 m deep, 1.0 m wide, and 1.0 m high. They are closed by wooden doors, which are normally unpainted and engraved with abstract motifs of the same style as *erong* sarcophagi, as well as *pa'tedong* buffalo heads. More recent tombs (style 2) are larger. Their doors measure *c.* 1.0 × 1.2 m, and their entrance areas are elaborately carved within the rock, creating a large recess space protecting the wooden door from rain, as well as a stone lintel, sill and jambs framing the wooden door. The latter is normally highly decorated with engravings and painted colours representing various symbols (buffalo heads, sun-motifs, spiral patterns), and some tombs even feature reliefs sculpted on the solid rock around the edges of the entrance recess (Fig. 2.13). The chamber too is larger, measuring *c.* 2.0 × 2.0 × 2.0 m. Locational patterns also vary across the two styles: older tombs tend to be in more discrete places, and at higher parts of the rock faces, while more recent ones are often found in more visible and accessible rock surfaces.

Unlike *erong* sarcophagi, *liang pa'* tombs have never been dated with radiocarbon analyses. It is therefore difficult to assess how style 1 and style 2 tombs relate chronologically. It is likely that their chronological development overlapped and varied locally, without a clear-cut transitional date all over Sa'dan Toraja. We have seen above that the first generation (style 1) of *liang pa'* were first constructed around the 17th century. One can roughly estimate the emergence of the second style (with more elaborate carvings, larger size, and more publicly visible locations) in the 1970s. This estimation is based on various lines of evidence, ranging from discussions with

Figure 2.14: A very old liang pa' *rock-cut tomb (Parinding boulder 10, Sesean). The tomb was cut immediately over a rock-shelter covering older* erong *sarcophagi (photo: G. Robin).*

Figure 2.15: Two generations of liang pa' *rock-cut tombs on boulder 21 at Bori' (Sesean): earliest tombs have small unpainted doors, while most recent tombs have large entrances with painted doors (photo: G. Robin).*

stoneworkers responsible for cutting tombs, to the dates occasionally inscribed on tomb doors, as well as old photographs of cemeteries showing their previous development stages at specific dates.

Inscribing dates and names on the doors of tombs seems to be a recent practice in Tana Toraja. Such inscriptions are found mainly on recent tombs (style 2). A single date often corresponds to the year in which the tomb was created and is accompanied by the name of the sponsor (*e.g., milik M. Tarukallo*, 'belongs to M. Tarukallo'; *dipahat K. Ne' Saro*, 'chiseled by K. Ne' Saro"; *D.L. Pasalli' dibuat*, 'made by D.L. Pasalli").[7] More rarely, a couple of dates can be inscribed, which refer to the dates of birth (*lahir*, 'born') and death (*meninggal* or *wafat*, 'died') of a person buried in the tomb. During our survey in 2017 we recorded inscribed dates on 55 tombs: tomb foundation dates (or individual death dates) from 1970 onwards all belonged to style 2 tombs, while the few dates in the 1950s–1960s were found on tombs which seem to be of an intermediary style, with a larger entrance than style 1 tombs but lacking of the size and elaboration of style 2 tombs. No date inscriptions were noted on style 1 tombs.

The ethnographic literature sometimes includes photographs of *liang pa'* cemeteries that were taken in the late 1970s (*e.g.*, Lo'ko' Mata in Koubi 2010: see Fig. 2.13; Lemo in Waterson 1990, 203). In these photographs, larger tombs with elaborately carved entrances (style 2), which can be seen today, are absent. This suggests that the more elaborate, style 2 tombs were created no earlier than in the 1980s.

The emergence and development of this new, distinctive style of tomb coincides with broader changes in Indonesia's recent history. President Suharto's *Orde Baru* ('New Order', 1966–1998) marked the end of a period of social, political and economic instability in the country and, to a lesser extent, in Tana Toraja. A period of economic development commenced in the 1970s, which was associated with local structural improvements, periodic economic migration of Torajans to other parts of Sulawesi and to other Indonesian islands and the beginnings of tourism in the Toraja highlands. In the 1980s, the development of new cash crops (cocoa and vanilla) in Sulawesi, including Tana Toraja, further fostered economic mobility and agricultural incomes. This resulted in a sharp increase in revenues and a boom in expenditures for funeral ceremonies (Waterson 2009, 117–118; de Jong 2013: see also Nooy-Palm 1986, 304). This sudden economic growth likely accounts for the emergence of a new generation of *liang pa'* tombs in Tana Toraja, with larger chambers and conspicuous entrances and decorations. This new form of social display would have suited both traditional noble families and *nouveaux riches*, *i.e.*, individuals with commoner or slave ascendance who benefited from the new economic opportunities and who used this new wealth as a way to enhance their social rank. The increased investments in rituals associated with death may also be attributable to an overall de-emphasis on fertility-related rituals due to prohibitions against Christians participating in such ritual practices, instituted by Dutch missionaries in the early 20th century (van der Veen and van der Veen 2023, 19).

In addition to the effects of the New Order economic boom, the new generation of *liang pa'* tombs could also be attributed to the identity politics of ethnic expression in Tana Toraja. Beginning in the 1970s, there was indeed a marked increase in expressions of ethnic identity that have been viewed as a reaction to violent attacks from Muslim paramilitary forces from the lowlands in the 1950s and 1960s, when there were attempts to form a separate Islamic state in South Sulawesi. The identity politics of Toraja cultural expression were further amplified by the increase in tourism, both foreign and domestic, to Tana Toraja beginning in the 1980s (Adams 2006, 39).

Table 2.1. Proportion of old and recent liang pa' *rock-cut tombs in study area A.*

	No. boulders or cliffs	No. of liang pa'
Old *liang pa'* (style 1) 1700s-1960s	37	189
New *liang pa'* (style 2) 1970s-today	160	380

In terms of spatial distribution, old and recent *liang pa'* are found in the same areas, at times in the same rock cliffs or boulders. If one examines the distribution of boulders and cliffs with old and recent *liang pa'* in the Sesean area to the north of Rantepao (Figs 2.16; 2.17), one sees that both styles have roughly the same widespread distribution. Their proportion, however, is significantly different: 37 cliffs/boulders have old tombs (style 1), representing a total of 189 *liang pa'* created between c. 1700 and the 1960s, while 160 cliffs/boulders have recent tombs (style 2), representing a total of 380 *liang pa'* built between the 1970s and today (319 of which were used, while 61 were newly built, or in the process of being built, and vacant in 2017) (Table 2.1). These figures exemplify the intensification of *liang pa'* tomb building resulting from the economic growth that started in the 1970s.

The districts on the slope of Mount Sesean (study area A) constitute one of the most dynamic regions for rock tomb cutting and use in Tana Toraja. The area is known for its long-standing stone cutting tradition; most of the stone workers specialising in tomb cutting all over Tana Toraja are from there (see Chapter 4). Menhirs (*simbuang batu*) are often quarried in that particular area, before being transported to different locations (Adams and Robin 2022). It is therefore not surprising to find so many tombs in the Sesean area itself. In areas located to the south of Rantepao (areas B and C: Fig. 2.16), however, tomb clusters are fewer and one can observe a flipped pattern, with older tombs found in higher numbers compared to recent ones. The cemetery of Lemo (district of Makale Utara), for instance, includes 66 old *liang pa'* and 14 recent ones (five of which were being cut during our visit in June 2017). At the cemetery of Batu Lappa' (Buri', district of Rembon) we counted 13 old tombs and one recent tomb. Finally, the cemetery of Salu Liang (Kole Sawangan, district of Malimbon Balepe') includes approximately 70 old tombs (Waterson 1995, 210), but only two recent ones were seen there during our visit. In these areas south of Rantepao, the lack of recent rock tombs can be explained not only by the lack of available rock surfaces (unlike in the mountainous Sesean districts), but also by the recent development of concrete house-tombs (*patane*) that are nowadays often preferred over traditional *liang pa'* monuments.

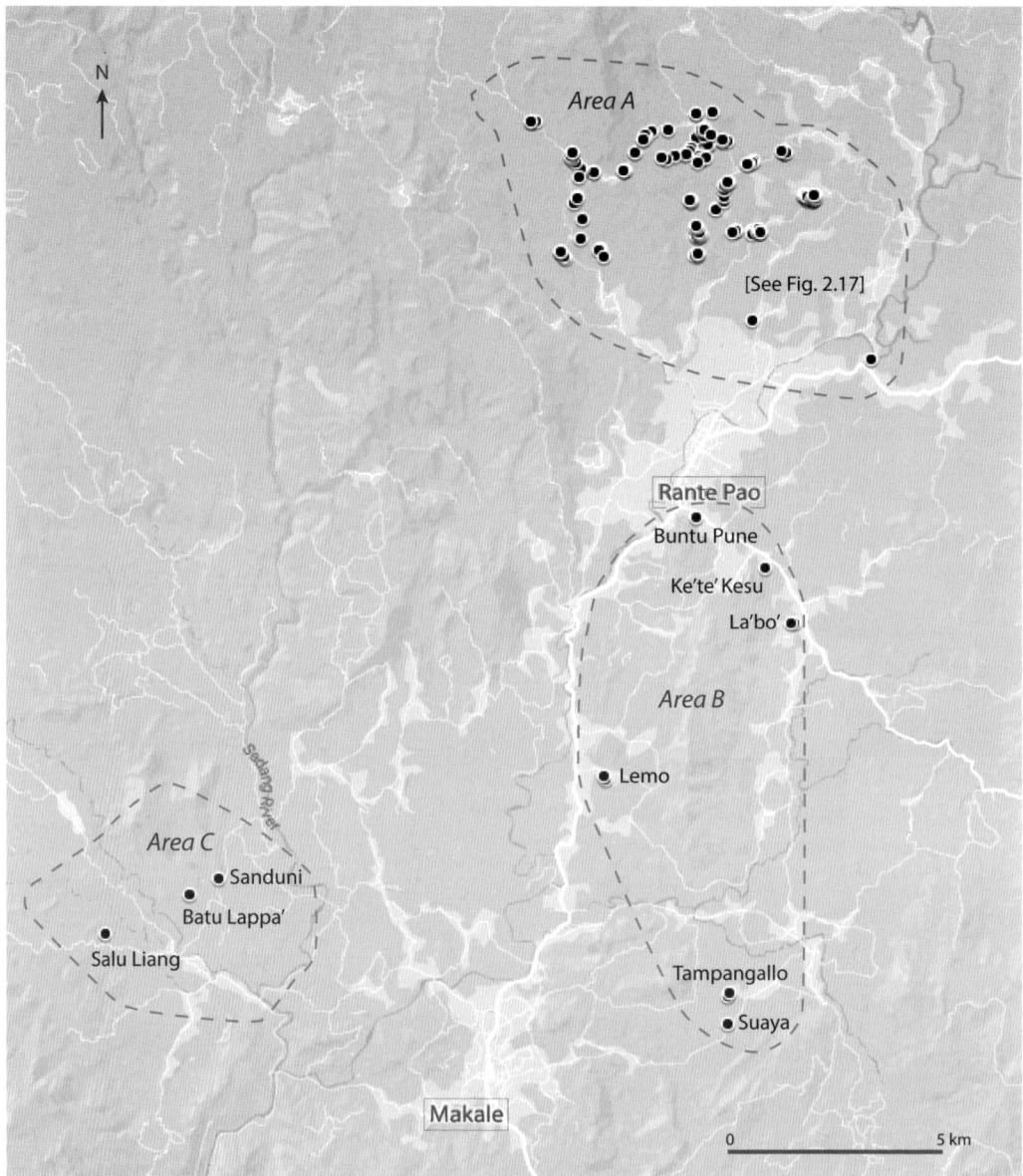

Figure 2.16: Location of rock-cut tomb cemeteries examined by the authors in 2017 in the Sa'dan region of Tana Toraja (map: G. Robin; basemap: Google).

House-tombs (*patane*)

This type of aristocratic burial is certainly the most diffused one over the entire Toraja country: unlike rock-cut tombs, house-tombs are found virtually everywhere today in the countryside in both Mamasa and Sa'dan regions. It is also the most diversified

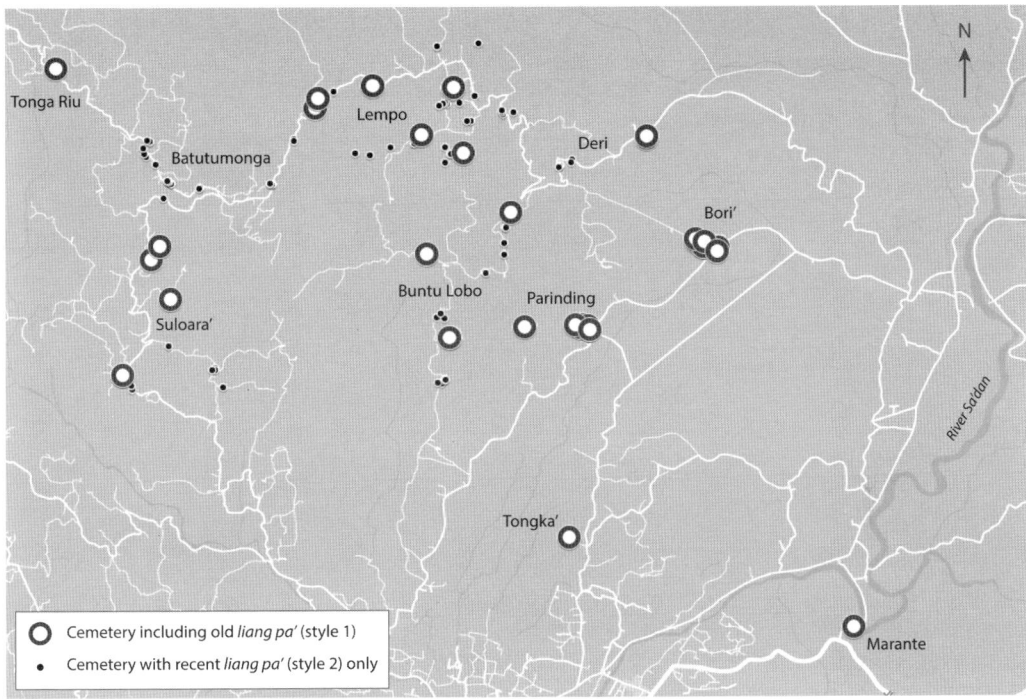

Figure 2.17: Location of rock-cut tomb cemeteries examined by the authors in 2017 in the Sesean districts (Area A) (map: G. Robin; basemap: Google).

category of tombs: under the broad label of 'house-tombs', we place different forms of structures, made with different materials, and in which burials are disposed of in different ways. However, all these monuments share the same concept of replicating a house in smaller form, which is used as a shelter for multiple burials.

Based on ethnographic literature[8] and our own field notes, two main generations of house-tombs can be identified in Tana Toraja. The two generations overlap chronologically and spatially. Like the two generations of *liang pa'*, they represent an evolution in the use of aristocratic tombs.

The emergence of house-tombs in Tana Toraja cannot be dated with precision, but they seem to pre-date the *liang pa'* tradition and were used in parallel to *erong* (Kruyt 1924, 167) (Fig. 2.18). The most ancient versions of house-tombs are found in areas that lack vertical rock faces (cliffs or boulders) in which to cut *liang pa'* or to hide wooden sarcophagi. Their origin, therefore, may be interpreted as a form of cultural adaptation to environmental conditions. In the Mamasa region, ancient house-tombs are called *tangdan* (or *tangngan*) or *batutu*. In the Sa'dan region, rock cliffs are more numerous and therefore ancient house-tombs are relatively rare, however a few were built in the agricultural plains around the capital town of Rantepao (Fig. 2.19), or in the eastern district of Buntao'. In this region, they are

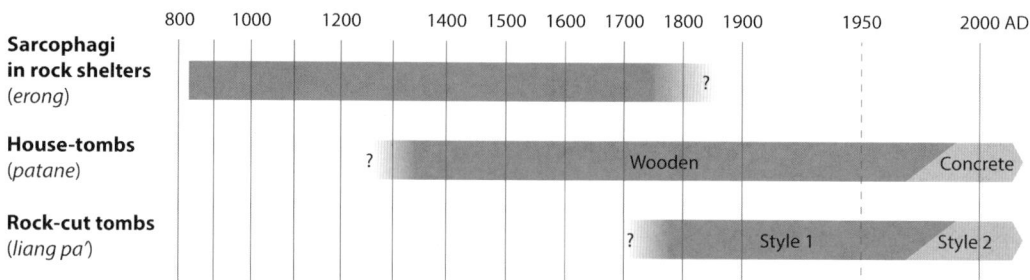

Figure 2.18: Development over time of the three main types of aristocratic burial in Tana Toraja (drawing: G. Robin).

Figure 2.19: Small-scale wooden house replica on top of a patane *tomb dug into a flat outcrop in the rice paddies of Kalambe (Tikala) c. 1930 (unknown author, Collection Wereldmuseum Coll.nr. TM-FV-0024-51).*

called *patane*. The house-tombs can be isolated or grouped together to form a cemetery of up to seven monuments.

Ancient house-tombs are located on hilltops. They consist of a wooden structure built over a large flat rock outcrop. The rock surface under the wooden structure is the receptacle for the burials. Bodies (wrapped in cloths) are either laid over the rocky floor, or placed into a vault that was hewn out from the top surface of the rock and covered by a wooden lid.[9] In the Mamasa region, where rocks are less common, bodies were not placed in rock-hewn vaults but inside wooden sarcophagi, grouped together and sheltered under wooden tomb-houses (Duli 2014). According to Kruyt (1924, 164–165), when the house-tomb was full, new corpses were placed in individual log coffins held outside the house-tomb under its protruding roof.

The wooden house structure standing over the burials can have different forms. It can be an elaborate imitation of a *tongkonan* smaller in scale, with its curved projected roof, posts, beams and wood carvings (Fig. 2.19). Such structures were usually open, which permitted the visibility of the objects that were deposited with the dead. Indeed, the wooden house was aimed not only to protect the burials themselves but also the belongings of the deceased (*e.g.*, hats), offerings (betel baskets and pouches) and wooden human effigies (*tau-tau*). These effigies, discussed in Chapter 5, represent the deceased individuals deposited in the tomb and are created on the occasion of funeral ceremonies of the highest rank only. The frequently reported presence of *tau-tau* in ancient *patane* attests to the strictly aristocratic status of such tombs in the past. According to Kruyt (1924, 167), *patane* were even restricted to the higher nobility (*puang*), while other nobles were buried in *erong* stored in rock shelters.

Some ancient *patane* are less elaborate, such as the tomb of Pangkaro in Tembamba (Buntao'). The monument is more than a hundred years old and was still standing in 2017, although in a poor state of preservation (Fig. 2.20). It consists of a simple closed wooden house with a flat thatched roof. This collective tomb was not used by a single *tongkonan* but shared by several in the area. Bodies were placed directly on the surface of the rock inside the closed 'house' structure. The door of the 'house' faces south (the location of Puya, the world of the dead), unlike actual houses, which have their entrance to the north. At Pangkaro, *tau-tau* effigies used to be displayed outside the *patane*, against its wooden walls, but they have now been removed to protect them from theft. The walls and beams of the *patane* are decorated with wood carvings (see Fig. 2.21 for a more recent example).

The more recent generation of house-tombs are also called *patane* in the Sa'dan region. They are built with modern construction techniques and materials (concrete, bricks, metal, *etc.*), and are often very simple in form (Fig. 2.22). They are smaller than the ancient *patane*, for they are often used independently by a single family, rather than being shared by an entire kin group. Unlike the ancient wooden house-tombs, these are found everywhere in the Toraja countryside, including in areas with a vivid *liang pa'* tradition such as Sesean. According to Roxana Waterson (1995, 207),

Figure 2.20: The old wooden patane of Pangkaro in Tembamba (Buntao'), today protected under a corrugated roof shelter (photo: G. Robin).

Figure 2.21: *A traditional wooden* patane *of recent facture in the cemetery of Buntu Pune (Kesu) (photo: G. Robin).*

the development and multiplication of these modern *patane* started in the mid-1980s. Like the recent style of *liang pa'* described above, their development can probably be associated with the economic growth of the New Order era in Indonesia. Indeed, it is important to highlight that modern *patane* can be built and used by any family who can afford it, and do not require an aristocratic background or an affiliation with a *tongkonan* (Table 2.2). They represent an example of the democratisation of monumental burial in Tana Toraja, as well as of a certain distancing from traditional practices and prohibitions, partly due to the influence of Christianisation since the early 20th century.

Modern *patane* are therefore very popular among Toraja families of lower-rank background, who have recently become wealthy and are keen to display their new status. Nevertheless, concrete tombs are also popular among old aristocratic families, either because the *patane* are being constructed in places where they have long been used traditionally since ancient times (*e.g.*, in Kesu), or because they offer advantages over *liang pa'* rock-cut tombs in terms of space and accessibility. Although *liang pa'* are still considered the most prestigious type of tomb in Tana Toraja today, the success of concrete *patane* can be explained by a combination of practical, economical and religious factors. In some areas, *liang pa'* cemeteries are full and contain little to no available rock surfaces for cutting new tombs. Since family tombs themselves are

Figure 2.22: Modern concrete patane *in La'bo' (Sanggalangi) (photo: G. Robin).*

Table 2.2. Main differences between ancient and recent patane house-tombs.

	Ancient house-tombs (up to 1980s)	Recent house-tombs (1980s onward)
Distribution	Mamasa mainly, rare in Sa'dan	Everywhere
Name	Tangdan (or tangngan), batutu, alang-alang, palakka, patane	Patane
Landscape setting	Hilltops	Anywhere
Orientation	Entrance faces South	Any orientation
Construction material	Wood, thatch, stone	Concrete, bricks, glass, metal
Burial type	Bodies wrapped in cloths (mebalun) & disposed in rock-cut pit under the structure or in wooden sarcophagi under the structure	Bodies in individual coffins, placed inside the structure
Rank	Higher nobles only	Anybody with enough wealth
Tau-tau effigies	Yes	No

running out of space, aristocratic families may decide to build a concrete *patane* instead of cutting a new *liang pa'*. Several modern *patane* are indeed built right at the bottom of old *liang pa'* cemetery cliffs, where they are 'taking over' from ancient rock graves. At several sites we visited, informants have described how concrete *patane* are sometimes built directly against the rock face and the entrance of an old family *liang pa'* in order to create a modern extension for the deposition of recent and future burials.

Moreover, modern *patane* can be placed anywhere in the landscape, in any land parcel belonging to the family. In terms of accessibility, this represents a practical advantage over *liang pa'* cemeteries, especially old ones which are often located in distant, remote areas. One of our informants in Buri' (Rembon) was affiliated to a *tongkonan* whose rock grave is in the old cemetery of Batu Lappa' (Fig. 2.16). The grave is located 600 m away from the *tongkonan* and is only accessible by way of a tiny path that extends across a series of rice paddies. A member of the family created a concrete *patane* several years ago, conveniently located on the edge of a road 100 m from the *tongkonan*. Although the old family *liang pa'* is not full, the family has preferred to use the modern *patane* instead in recent funerals, since it is more accessible and also easier to use as a burial structure. Bodies are typically deposited with individual coffins inside the spacious chamber of the *patane*, while using an old rock grave requires carrying a wrapped body (see Chapter 5) up a cliff face and inserting it through the small opening of the chamber.

Bodies of deceased nobles often need to be kept several months inside *tongkonan* houses before funeral ceremonies take place. Today, bodies are often embalmed by injections of formaldehyde. This chemical agent preserves the body well in the short term but results in faster deterioration of the corpse in the longer term. In this case, a *patane* offers another advantage, as it has more space thereby making it less likely that a corpse would get crushed, as would be the case if the body were to be interred in a crowded *liang pa'*.

Building a concrete *patane* is cheaper than cutting a *liang pa'*. In 2017, it cost 40–50,000,000 Rupiah (*c.* 1800–2250 GBP/2400–3000 USD) to have a large *patane* built, while a *liang pa'* cost 60–100,000,000 Rupiah (2700–4500 GBP/3600–5950 USD) (see Chapter 4). However, this cost differential may not be a factor for aristocratic Toraja families who have access to available rock faces. Finally, religious changes may be another reason for preferring modern *patane* over older *liang pa'*. According to Roxana Waterson (1995, 207), Christian Torajans, particularly fundamentalists, sometimes feel uncomfortable with the idea of lying in the *liang pa'* together with their pagan ancestors.

Despite the success and exponential development of concrete *patane* in Tana Toraja in recent decades, *liang pa'* rock-cut tombs have not seen their prestige diminished. They are still regarded as the most enduring monuments and as the most appropriate burial for families of aristocratic ranks.

Lower rank burials

Caves were used in the past as places for collective burials for commoners (Grubauer 1913, 203; Duli *et al.* 2019, 4). Such places are called *lo'ko'* ('hole', 'cave', 'gallery': Koubi 1982, 195). Bodies of all ages and sexes were wrapped in cloths and piled up with no other form of identity marker. This tradition has disappeared today, although the deposition of individual rectangular coffins in communal caves at the foot of *erong* cemetery cliffs at Londa and Ke'te' Kesu was a common practice for commoner class burials until recently.

In-ground burials (or 'flat graves' in archaeological terms) are not uncommon in Tana Toraja. In the past, they were the only option for slaves, commoners and certain lower nobles. According to Kruyt (1924, 165), bodies were placed in supine position in a grave, which, once back-filled, was only marked on the ground by two stones, one at the head and the other at the foot end. Nowadays, in-ground burials are part of recent developments associated with religious conversions. Most of those we could observe in 2017 were explicitly Muslim or Christian graves. They did not express any visible connection, such as decorative motifs, to traditional Toraja cultural traditions.

Infant burials

The forms of burials described above are for adults and children. Infants from noble families receive a different funerary treatment. Stillborn infants were traditionally placed in pots and buried on the west side of the *tongkonan* house (Kruyt 1924, 138). Interestingly, this practice can be contrasted with that following successful births, which involves burying the placenta on the *east* side of the house (Waterson 2009, 185).

This can be linked to the east (life) *vs* west (death) opposition that structures Toraja rituals (Nooy-Palm 1986).

Infants who have died after their birth, but before growing teeth, are buried in small individual spaces that are hewn out from large living trees (Fig. 2.23). These small 'tree-cut tombs' are called *liang piah* (Nobele 1926, 57; van der Veen 1940, 307) or *disilli'* (Nooy-Palm 1986, 183–184). Only special trees are selected for such burials, in particular, the *kapok* tree (*Ceiba pentendra*) or the *lamba* tree (or banyan tree, *Ficus elastica* or *Ficus religiosa*). The latter has a special ritual status in Tana Toraja and can live for hundreds of years (Nooy-Palm 1979, 219; Ismail and Yusuf 2019). The distinctive white sap of these trees is regarded as a form of natural milk, replacing the mother's breast milk. It is believed that the babies will grow along with the tree after the burial, feeding on the tree.

Bodies of infants are wrapped in cloths and placed upright in the small vertical cavity, which is then closed by

Figure 2.23: Liang piah *infant burials in Kambira (Sangalla). Small individual burial spaces are cut into living trees and closed by palm fibre mats (photo: G. Robin).*

a palm fibre mat. The mat is fixed on the tree with a specific number of sticks nailed into the trunk: six for the higher nobles (*tana' bulaan*) and four for lower nobles (*tana' bassi*). After several years the palm fibre mat decays, the tree heals with bark covering the hole, which only leaves a rectangular scar on the tree trunk. Up to 25 individuals can be interred in a single tree. A ceremony is held for each burial, which involves the killing of pigs, as for the funeral of an adult noble, except that the meat of the sacrificed pig is not shared among the attendees and is prohibited from being taken away. This type of burial is still practised today.

Notes
1. The term 'coffin' is often used to describe these old burial containers (*e.g.*, Nooy-Palm 1979; Waterson 1988; Duli 2014; 2015). We think that the term 'sarcophagus' (Crystal 1985; Nooy-Palm 1999) is more appropriate, considering that they were intended to be displayed rather than buried and to be used for more than one individual interment.
2. *Erong* is the most common term among Sa'dan Toraja to refer specifically to these cliff sarcophagi. Other terms can occasionally be found in the ethnographic literature; they are used in specific

areas only, and present some ambiguity as they can be used to refer both to ancient wooden sarcophagi and to more recent wooden coffins still used in certain funeral ceremonies. Such terms include: *rapasan* (in Rantepao), *kayu mate* (in Sangalla'), *mandu* (in Duri), *bangka* (in Rante Bala), *karopi* (near Mebali) and *balombong* (in Bittuang) (Kruyt 1924, 365–366; Keers 1939; Nooy-Palm 1979, 196; Koubi 1982, 101).

3. The dramatically curved roofs of aristocratic Toraja origin-houses (*tongkonan*) are often described as boat-shaped, a design whose origin may have mythological signification (Waterson 1988, 49). The *erong* sarcophagi, as imitations of these houses, are therefore sometimes described as boat-shaped since their incurved lid is a copy of the houses' roof.

4. In the region of Mamasa, ancient sarcophagi are given different names according to their shape: boat-shaped (*bangka-bangka*), buffalo-shaped (*tedong-tedong*), horse-shaped (*narang*) and round-shaped (*talukun*) (Duli 2014).

5. Social memory of *erong* interments may still exist in a limited number of cases. We were told that the latest *erong* burials that took place at the cemetery of Tampangallo (Sangalla) were associated with *tongkonan* Sarapun whose members were still able to identify their family sarcophagi. At the cemetery of Ke'te' Kesu (Kesu), we observed an *erong* sarcophagus that was made in 2001. It was commissioned by a high ranking *tongkonan* specifically to collect bones of their ancestors from an older, decaying *erong*. *Erong* burials are no longer practised today and this exceptional occurrence can be regarded as an act of relocation rather than a proper burial.

6. Songi Pataki may be 'Songgi Patalo of Lemo' mentioned by Roxana Waterson (2009, 51).

7. A single date with no other inscription can be misleading, as it does not necessarily refer to the foundation of the tomb, but can also refer to a burial event, which may have happened several generations after the creation of the tomb. For instance, the most recent grave at the cemetery of Batu Lappa' in Buri' (Rembon) has a '1982' painted underneath the door. According to our informant, who is related to the cemetery, the tomb was actually founded by his grandparents more than a hundred years ago, and the 1982 date corresponds to the last burial (his uncle) that took place in the tomb. It is therefore better to have a date together with a sponsor's name to ascertain the creation date of a tomb.

8. Primary references dealing with Toraja house-tombs are: Kruyt 1924, 163–168; Nooy-Palm 1979, 260; Brisbois and Douvier 1980, 116–117; Koubi 1982, 195, 229–230; Waterson 1995, 207; 2009, 178–179; Duli 2014; Duli *et al.* 2019.

9. The word *patane* in van der Veen's Toraja-Dutch dictionary is defined as 'a great stone into which a hole has been cut downwards; in this hole the significant [noble] dead are buried; as a cover, a house is placed on top, which has the shape of *e.g.* a Toraja house, or a square shape with a pyramid-shaped roof [...]; in that house, the *tau-tau*, a wooden doll, the image of the dead, is deposited' (van der Veen 1940, 451: Robin's translation). This definition suggests that the word *patane* (in the Sa'dan region) referred primarily to the rock-hewn vault, not the house structure over it.

Chapter 3

The anatomy and decoration of *liang pa'*

This chapter gives a detailed description of the main components of *liang pa'* rock-cut tombs from both the old and recent generations (style 1 and style 2, defined in Chapter 2). We start by discussing the architectural design of the tombs, from the burial chambers to the entrance areas. Then, we present the main forms of closing systems that are added onto the solid rock architecture in order to secure their burial contents. Finally, we discuss the traditional engraved and painted decorations executed on the doors, as well as the more recent practice of sculpting motifs around the outer rocky entrances of the monuments.

These elements represent the 'fixed' (or structural) part of the monuments. The 'mobile' or more dynamic elements, such as the burials placed inside the tombs and the ritual depositions in front of them, will be discussed in Chapter 5.

Burial chamber

The architecture of *liang pa'* rock-cut tombs is overall very simple. It has two main parts. The first is a short, recessed entrance area, cut into the rock face, which provides a shelter for the offerings to the dead and for the wooden door. Behind the door is the second part of the tomb: a larger cubic space which forms the burial chamber. The size of the burial chamber varies from one tomb to another, depending on the period of construction (older *liang pa'* are smaller than recent ones: Fig. 3.1) and on the degree of expenditure, which can be quite high, that the sponsor is able to invest in its construction (see Chapter 4). In any case, the burial chamber must be large enough to receive multiple inhumations, *i.e.*, bodies laid out in an extended supine position (see Chapter 5).

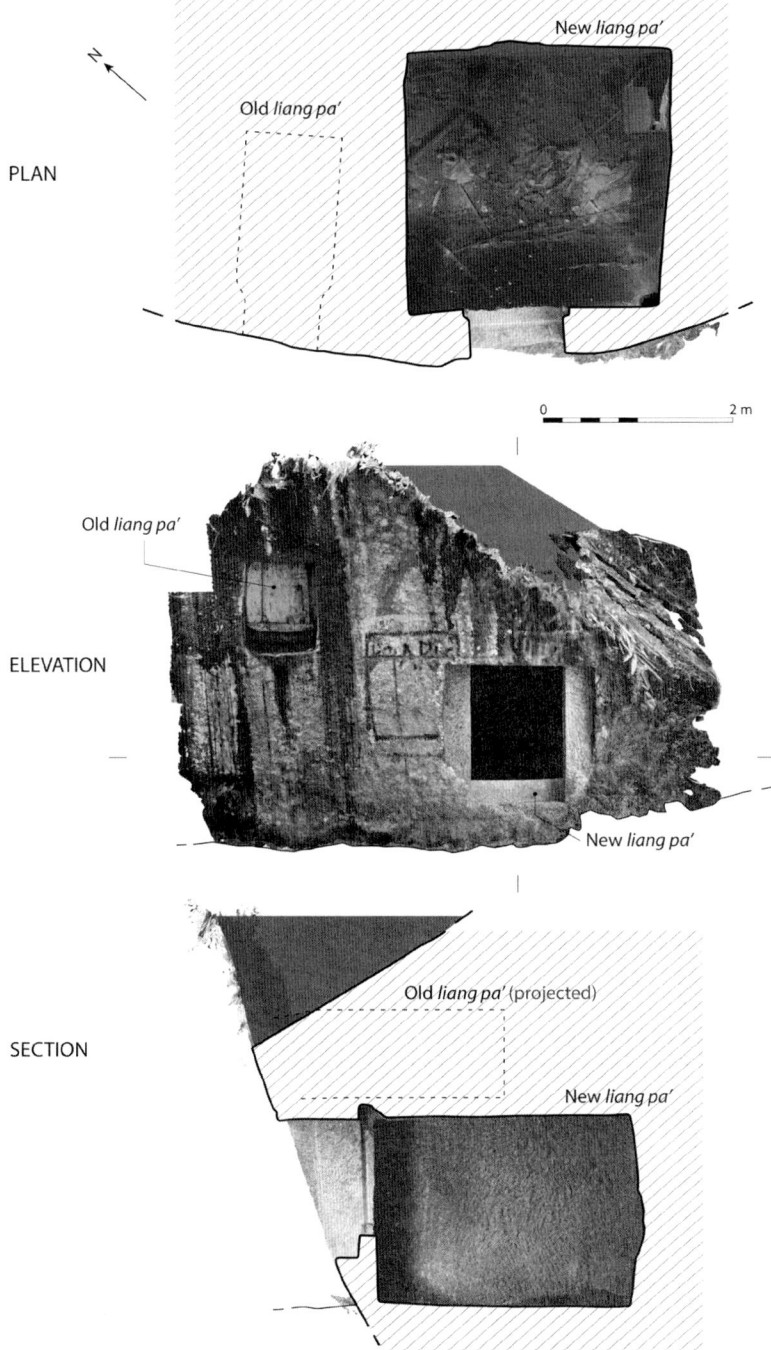

Figure 3.1: Western face of boulder 4 in Deri (Sesean), with an old liang pa' (style 1) and a new liang pa' (style 2). The latter was recently completed and not inaugurated yet at the time of our survey (June 2017) (images: G. Robin).

Old tombs (style 1)

The smallest chambers are found in *liang pa'* belonging to the older generation (Fig. 3.1: Pl. 7). We measured one of them (disused) at Salu Liang as 0.80 m high, 1.10 m wide, 1.60 m deep. Its entrance opening was 80 × 50 cm. Most old tombs are still closed or are located in high, inaccessible positions in cliffs, which means we could not carry out multiple measurements. However, various sources confirm these figures. According to one of our guides, old chambers are *c.* 2.0 m deep, 1.5 m wide, and 1.5 m high. According to Albert Grubauer, who observed a *liang pa'* in the process of being cut in 1911, the chamber was 2 m deep and 1 m high (Grubauer 1913, 215–216; Brisbois and Douvier 1980, 116). According to a senior stone carver we met in Bori', the chambers' depths were typically 1.80 m, while their widths had to correspond to the distance between the carver's two elbows stretched out sideways. These chambers were designed to receive multiple body depositions but in some cemetery sites, such as Lemo, we were told that certain *liang pa'* were cut to receive only one body and were therefore smaller.

A few old *liang pa'* present a different design, with the rock chamber being laid parallel to the rock face instead of perpendicularly (*e.g.*, in Lo'ko' Mata, see Fig. 2.13). These tombs have a wide rock opening (closed with a wooden panel) which corresponds to the width of the chamber. In these tombs, bodies are placed parallel to the rock face. The reason for this particular design is unclear, but it may have been easier to cut compared to the perpendicularly cut tombs, and therefore cheaper to commission.

Recent tombs (style 2)

The size of the burial chambers inside recent *liang pa'* is easier to evaluate, since these tombs are still being hewn nowadays and are easily accessible (Figs 3.1–3.3, see also Fig. 1.2). Many newly-cut tombs remain vacant and open until the death of their sponsor, which often happens several years after the tombs are complete. This gave us an opportunity to make direct measurements and examinations inside several tombs from various areas. During our 2017 field survey, we observed a total of 411 recent tombs (style 2), of which 345 were in use, while 52 were complete and vacant, and 14 were in the process of being cut. Table 3.1 shows measurements from 21 of them. According to local stone workers who create these tombs, contemporary *liang pa'* vary slightly in size, depending on the costs that the sponsors are ready to pay (see Chapter 4), but the smaller ones have to be at least 2 m deep, while the largest can be up to 4 m deep. While ancient *liang pa'* received bodies wrapped in cloths (*mebalun*), recent ones are designed to receive solid coffins. The latter are typically 2 m long, therefore the standard depth of *liang pa'* nowadays is specifically set to 2.20 m in order to allow enough space to fit these standard coffins. The heights and widths of the tombs are variables that are determined according to the will and wealth of the sponsors.

Recent tombs are therefore much larger than *liang pa'* of the older generation (style 1). They have more space to receive several large coffins and have a sufficient interior height to allow people to enter and stand inside the chamber to sort out the placement of new coffin depositions, while in the case of old tombs, wrapped

Table 3.1. Measurements from 21 recent liang pa' (style 2).

Village	Cemetery	Depth (m)	Width (m)	Height (m)	Volume (m³)	Addit. space
Bori'	Boulder 21	3.00	4.00	2.00	24.00	–
Buntu Lobo	Boulder 1	2.20	3.00	2.00	13.20	–
Buntu Lobo	Boulder 6	2.00	3.00	2.00	12.00	–
Buntu Lobo	Boulder 8	2.20	4.00	3.00	26.40	Back recess
Buntu Lobo	Boulder 13	1.90	2.00	2.00	7.60	–
Buntu Lobo	Boulder 16	2.00	2.00	2.00	8.00	–
Buntu Lobo	Boulder 20	2.20	3.00	2.00	13.20	–
Buntu Lobo	Boulder 29	2.20	4.00	2.30	20.24	Side recess
Deri	Boulder 4	2.20	2.50	1.80	9.90	–
Lemo	Cliff B	2.00	4.00	2.00	16.00	Back recess
Lemo	Cliff B	2.20	3.00	2.00	13.20	–
Lempo	Boulder 17	2.30	3.00	2.50	17.25	–
Lempo	Boulder 17	2.20	3.00	2.00	13.20	–
Lempo	Boulder 17	2.00	2.00	2.00	8.00	–
Lempo	Boulder 18	2.20	3.00	2.20	14.52	–
Lempo	Boulder 18	2.20	2.00	2.00	8.80	–
Lempo	Boulder 18	2.20	3.00	2.00	13.20	–
Lempo	Boulder 26	2.50	2.30	2.50	14.37	Side recess
Parinding	Boulder 12	2.20	3.00	2.00	13.20	–
Suloara'	Boulder 10	2.50	2.30	2.30	13.22	–
Suloara'	Boulder 12	2.20	2.00	2.00	8.80	–

bodies were inserted through the small door opening from outside the chamber. The development of tomb cutting since the 1970s, and the increasing size of burial chambers, reflect the rise in economic revenues and ceremonial expenditures in Tana Toraja discussed in Chapter 2.

Special features
In the vast majority of cases, burial chambers consist of a simple cubic space. However, additional spaces or compartments can occasionally be created into the rock, off the cubic chamber. These can be described as upper recesses or benches, and are created to receive a single coffin. As stone workers explained to us, this separated space is prepared exclusively for the sponsor of the tomb, as a way of preventing their coffin from getting crushed under later coffin depositions (*i.e.*, the coffins of their descendants stacking up in the chamber over future generations). The recess ensures that the coffin will stay safe and be always positioned 'above' later burials. When such recesses are created, there are only one per tomb, and they are located either off the back or a side wall of the chamber (Fig. 3.3).

3. The anatomy and decoration of liang pa'

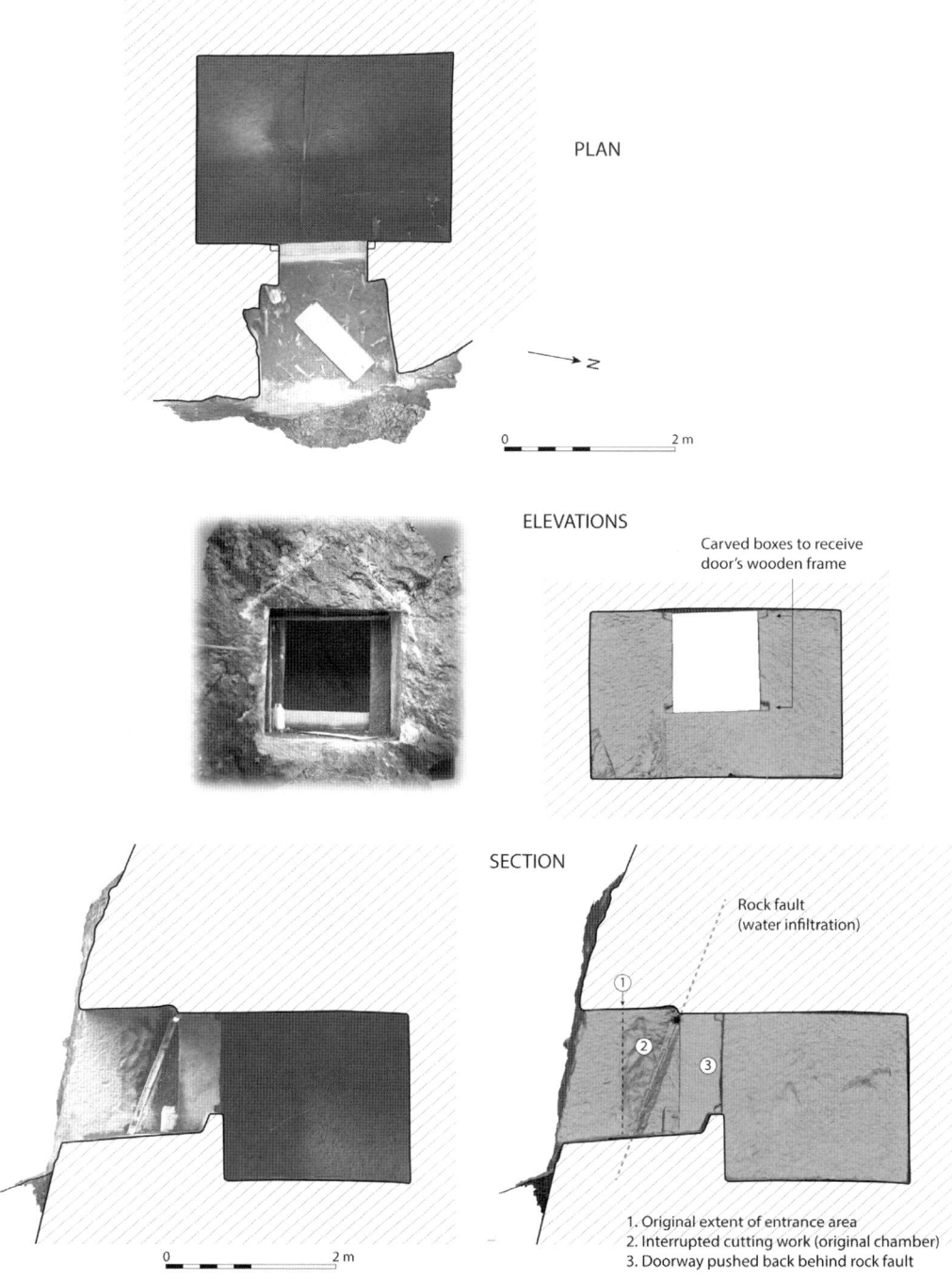

Figure 3.2: A newly hewn liang pa' at Lemo (Makale Utara). The unusual length of the entrance area is due to the presence of a rock fault which the masons had to avoid to create the burial chamber (images: G. Robin).

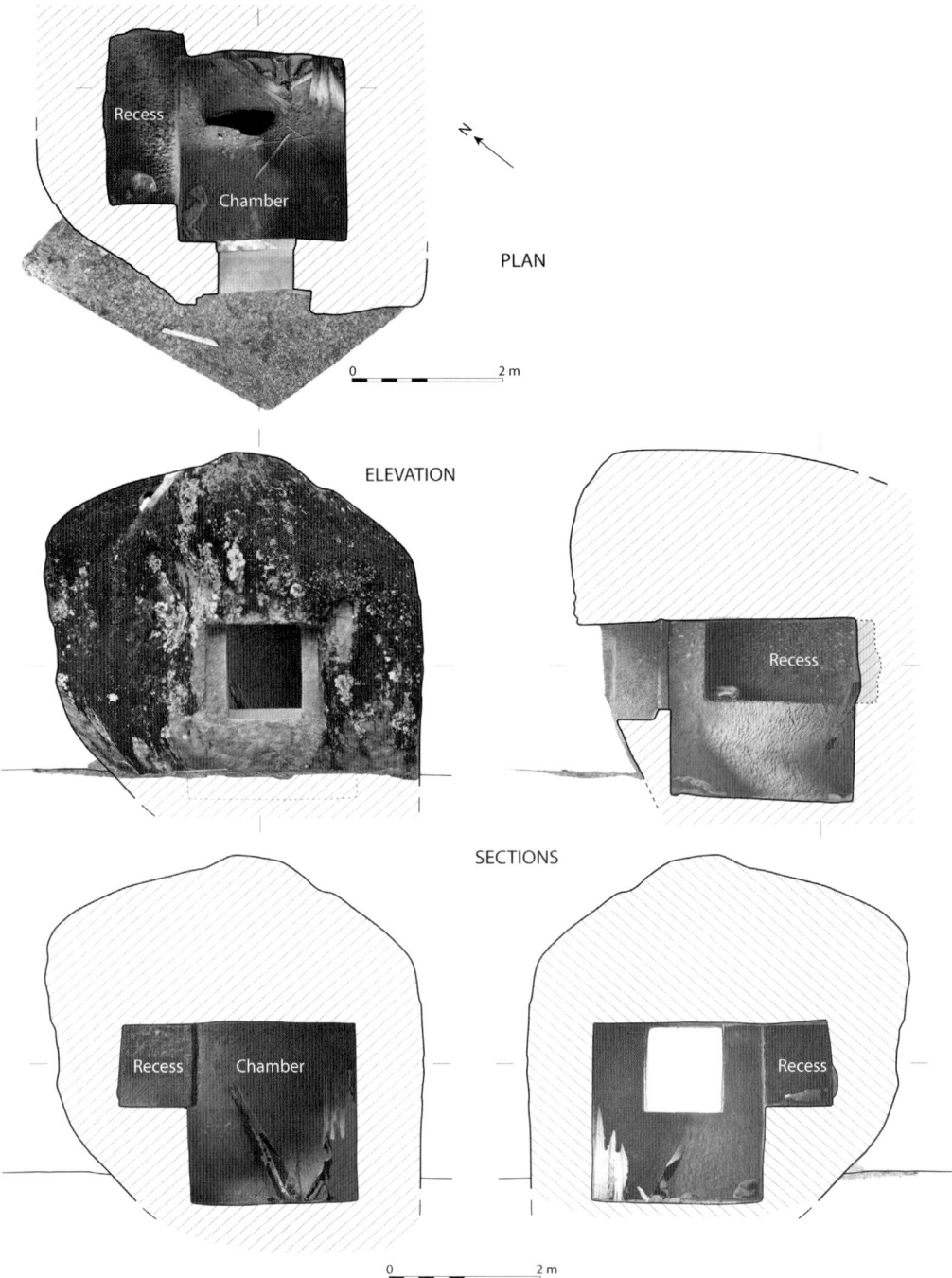

Figure 3.3: Boulder 26 at Lempo (Sesean Suloara'), with its recently hewn liang pa'. *The tomb's chamber has a separate recess space which is reserved for the coffin of the sponsor of the tomb. Coffins of the other members of the family will be deposited in the main chamber (images: G. Robin).*

According to our observations, coffin recesses are not very frequent in *liang pa'* generally. From an architectural design point of view, they represent the only *structural* means of compartmentalising the tomb space (although see Chapter 5 for other strategies of separating groups of bodies inside tomb chambers). Kruyt (1924, 162–163) and Nooy-Palm (1979, 260) mentioned that rock tombs of important nobles (presumably in the past) sometimes had an antechamber that preceded the main burial chamber. The antechamber, they stated, was used exclusively to bury the slaves of the *tongkonan* group so that they could keep serving their noble masters (buried in the second chamber) in the afterlife. During our 2017 investigation, we were informed of old burial practices involving slaves deposited together with their master in rock-cut tombs (see Chapter 5); however, we never encountered evidence for tombs with more than one chamber. Stone artisans we talked with were not aware of such elaborate tombs either. This suggests they were very rare and restricted to highly hierarchised contexts in Tana Toraja, such as the princely nobles (*puang*) from Sangalla.

On two occasions, we were informed of the existence of joint tombs, which could also be described as 'twin tombs'. At the cemetery of Salu Liang, an elderly woman pointed out two tombs located side by side at the top of a large boulder. She explained that a passageway connected the two tombs inside the boulder. One tomb was built by her grandfather, while the other one was built by her father. A senior stone artisan we met at the cemetery of Bori' told us that he had occasionally made tombs with a single wide chamber and two entrance openings side by side. He explained that these special tombs (which are very rare) were jointly commissioned by two families who had close connections and desired to share the same burial space. When viewing such a tomb from the outside, it looks as if each family has its own separate tomb (and each family actually uses its own entrance) but, in reality, the dead from both families share the same rock chamber.

Entrance area

The entrance to the burial chamber is closed by a wooden door, which is found at the back of a rectangular recessed space hewn out from the rock face. The depth of the recess varies typically 20–50 cm. The aim of the recessed entrance is twofold: to create a stone bench on which to place the various offerings from the living to the dead and to protect the wooden door from rainfall and water streaming down the rock face.

Protecting the chamber from humidity is an important concern in tomb design. At many cemetery sites, both ancient (Fig. 3.4) and recent (Figs 1.2; 3.2), the rock surface above tomb entrances was carved with inverted V-shaped grooves which are aimed at channelling rainwater away to the sides of the entrance. In recent years, concrete overhangs have sometimes been built over the entrance of tombs in order to create more protection (see Fig. 3.15). Water may occasionally, and unexpectedly, come from the inside of the rock itself. At a newly cut tomb we examined in Lemo, the design of the entrance had to be slightly modified to adjust to issues of water infiltrating from an internal fault (Fig. 3.2). During the cutting process, the carvers first completed the entrance recess and then, while they progressed deeper into the rock and started to hollow out the side walls to open up the chamber space, they found a natural sub-vertical fault in the rock. To prevent the fault from extending across the burial chamber (provoking water leaking onto the burials), they decided to extend the length of the recessed entrance, creating an unusually deep entrance (1.70 m), which enabled them to 'push' the burial chamber further back into the rock, beyond the fault plane.

The style of the entrance is quite different between older and more recent generations of *liang pa'*. Old tombs have a small and simple recess hewn out from the rock face, with an opening typically 0.50–0.75 m wide and 0.75–1.00 m high. In recent *liang pa'*, the entrance space is larger (typically 0.80–1.00 m wide and 1.00–1.20 m high) and displays a more elaborate design, with two sculpted side jambs and a low threshold framing the doorway (Figs 3.1–3.3).

Figure 3.4: Boulder 1 at Bori' (Sesean), with ancient liang pa' *rock-cut tombs (photo: G. Robin).*

Closing systems

Tombs are closed with a wooden (occasionally stone) door that is placed once the tomb's cutting work is completed (see Chapter 4). The door is located at the intersection between the entrance area and the burial chamber. It is composed of a frame and a shutter assembled together in a more-or-less complex system.

The way the entrance space and the burial chamber are articulated is notably different in old and recent generations of liang pa'. In old tombs, the height of the entrance corresponds to the height of the chamber, both forming a continuous tunnel (Pl. 7). In recent tombs, the doorway is located c. 1 m above the floor of the chamber. This may explain why wooden doors in old and recent tombs are constructed differently.

Liang pa' are collective burials that receive body depositions over generations, with human bodies piling up in the burial chamber. In old tombs, bodies are only wrapped in cloths (without coffins), which means oldest burials (at the bottom of the 'stratigraphy') become loose when they decay. These old burials need to be contained when one opens the tomb to insert new burials. This might explain why doors in old tombs include a fixed lower panel (under the small movable shutter) that acts as a barrier holding the burials (Figs 3.5 and 3.6). By contrast, recent tombs have larger chambers with floors located 1 m down from the entrance opening and are meant to prevent body parts from falling down outside of the tomb when it is opened up to receive new coffins. As a result, the door is a simpler assemblage of a regular frame with a large shutter.

Figure 3.5: Buffalo head motifs (pa'tedong) *carved on the wooden shutter of ancient* liang pa' *rock-cut tombs. Cemetery of Pana' (left: Tonga Riu, Sesean Suloara') and Buntu Lobo boulder 10 (right: Sesean) (photos: G. Robin).*

Figure 3.6: Left: basket lid motif (pa'kapu baka) carved on the wooden shutter of an ancient liang pa'. Right: ancient liang pa' with geometrical pa'erong motifs on the jamb and sill panels; the original carved wooden shutter was replaced by a more recent, plain one. Parinding boulders 11 and 10 (Sesean) (photos: G. Robin).

Only hardwood types are used to create tomb doors: *uru* (*Magnolia vrieseana*), *nangka* or jackfruit (*Artocarpus heterophyllus*), *sendana* or sandalwood (*Santalum album* L.), and *baringing* (*Ficus benjamina* L.). The same species of hardwood are used for other key ritual apparatuses, such as the *tau-tau* effigies of the dead (Crystal 1985, 131), certain pillars in *tongkonan* constructions (Nooy-Palm 1979, 241) and, in the past, *erong* sarcophagi (Duli *et al.* 2019).

Theft concerns relating to *liang pa'* were much greater in the past than in recent decades. Older *liang pa'* reputedly had elaborate locking systems embedded in their wooden door assemblages. According to Roxana Waterson, the door shutter is 'cunningly fastened by a concealed bolt on the inside' (Waterson 1988, 37). According to informants from the area of Salu Liang (where Waterson did most of her fieldwork), only certain *liang pa'* with stone shutters have this intricate closing system. In that particular cemetery, such doors are composed of three stone slabs (not wooden panels) assembled together and opening them requires a specialist who uses an iron stick to unlock the system. The use of stone as a material for tomb doors is quite rare. In our field visit in 2017 we noted stone doors in only two areas: in the cemetery of Salu Liang, and in Buntu Lobo (see Fig. 3.10, below), the latter area having a long-established tradition of stone cutting.

Doors with wooden shutters do not seem to be associated with particularly complex locking systems. Most shutters have a central handle (sometimes made out of a pair of buffalo horns: Fig. 3.4). The sill and header have grooves that hold the shutter, or perforated hinges, that allow the shutter to rotate (Fig. 3.7).

Decorations

Toraja *liang pa'* are decorated with carved and painted motifs, which nowadays contribute to their role as touristic attractions. Decoration is restricted to the exterior part of the tombs, *i.e.*, the entrance area. There are absolutely no carvings or paintings on the walls or the ceiling of the burial chamber. The tomb decorations, therefore, are not intended for the dead but for the living and the outside world. The decorations are mainly found on the wooden door that closes the tombs; however, a few recent tombs display motifs directly carved on the rock surface surrounding the entrance recess.

Figure 3.7: Tomb doors with sculpted human representations (tau-tau). *The jackfruit doors were made c. 70 years ago for* liang pa' *rock-cut tombs but were never sold and used. They are now part of a private collection in Tonga Riu (Suloara') (photo: G. Robin).*

The presence of decorative motifs on tomb doors has been noted by travellers and ethnographers (Grubauer 1913, 200; Wilcox 1949, 85; Nooy-Palm 1979, 259–260; Brisbois and Douvier 1980, 116); however, no studies had been devoted to Toraja tomb art before ours, involving a systematic analysis of the motifs, a discussion of their meaning and of their relationship with the architecture of the monuments. This gap contrasts with the excellent studies that were devoted to the decoration of *tongkonan* houses by Toraja scholars (Kadang 1960; Sande 1991) and western social anthropologists (Nooy-Palm 1979, 238–240; Waterson 1988; 1989). The social significance of Toraja tomb decoration and, in particular, the buffalo head motif, has recently been discussed by one of us as part of a literature-based review of tomb decoration in South-East Asia (Robin 2017). This work has highlighted the importance of the buffalo head motif and its multiple roles in both ritual beliefs (*e.g.*, protecting entrances from evil influences) and strategies of social display (as an active symbol of wealth and status). Here we would like to offer a more empirical take on Toraja tomb decoration, by discussing its character and diversity.

The discussion below is based on our systematic survey in the Sesean districts (study area A), and on visits to cemetery sites in the southern districts of Kesu, Sanggalangi, Sangalla, Makale Utara, Rembon and Malimbong Balepe' (areas B and C). During our systematic survey in the Sesean districts, we recorded 579 tombs. For the sake of the discussion on decoration specifically, we can exclude tombs that were in the process of being cut, and those recently completed but still vacant (which often have no door or just a temporary one). This gives us a total of 512 tombs, which we will use as a basis for the statistical figures presented in this chapter (see also Appendix).

Preservation of doors and decorations at cemetery sites

The first aspect to consider is the availability of the data, *i.e.*, the extent to which doors and their decoration are preserved and observable in the field. Unsurprisingly, older tombs are not as well preserved as recent ones. According to our survey, only a minority of old *liang pa'* still have their original, decorated door shutter. In most cases, the doors are either: absent as a result of natural decay or theft, with sometimes only the wooden frame remaining (sometimes also decorated with smaller secondary motifs; not discussed here); replaced by an expedient shutter (not decorated: Fig. 3.6); or not clearly visible enough to allow us to identify the motifs (tomb is too high up in cliff, too poorly preserved or obscured by vegetation). In total, the door survival rate in old tombs can be calculated to 37% (60 out of 164 tombs; Table 3.2). By contrast, recent *liang pa'* show much better preservation (90%: 314 out of 348 tombs; Table 3.3).

Decorated vs undecorated tombs

The second aspect to consider before discussing tomb decoration is the proportion of tombs that are deliberately undecorated. By this, we mean tombs with doors that are

Table 3.2. Preservation of doors and decorations in old (style 1) liang pa' rock-cut tombs.

Location	No. old liang pa'	Decoration visible	Decoration invisible	Original door replaced	No door
Batutumonga	1	–	–	–	1
Bori'	42	16	9	15	2
Buntu Lobo	4	1	–	2	1
Deri	3	1	–	–	2
Lempo	14	7	1	1	5
Marante	3	2	–	–	1
Parinding	29	6	5	13	5
Tonga Riu	13	11	–	2	–
Suloara'	44	15	1	10	18
Tallunglipu	11	1	–	–	10
Total	164	60	16	43	45

Table 3.3. Preservation of doors and decorations in recent (style 2) liang pa' rock-cut tombs.

Location	No. recent liang pa'	Decoration visible	Decoration invisible	Original door replaced	No door
Batutumonga	36	28	8	–	–
Bori'	52	43	9	–	–
Buntu Lobo	43	42	1	–	–
Deri	25	23	2	–	–
Lempo	47	40	7	–	–
Parinding	28	27	1	–	–
Tonga Riu	77	74	2	–	1
Suloara'	40	37	3	–	–
Total	348	314	33	0	1

not ornamented with traditional Toraja or Christian motifs, but simply left plain or painted all over with one or two colours (Fig. 3.9). Temporary or replacement doors are not included here. The survey reveals that undecorated doors are surprisingly common, and not only for recent tombs. They represent 16 out of 60 old *liang pa'* (27%), 58 out of 314 recent *liang pa'* (18%) and, overall, 74 out of 374 tombs (20%). Why would families decide to leave their tomb door undecorated? There are two main possible explanations. If the tomb owners are not from a noble background (*e.g.*, commoners with recent wealth), or from lower/poorer noble background, it would not be appropriate for them to display traditional motifs that are the exclusive marks of the Torajan nobility.[1] That is the main explanation that was given to us. Likewise, we were told that some *tongkonan* origin-houses belonging to commoners could not be decorated for the same reason (see also Waterson 1988, 34). There may be religious motivations too, nowadays, to some extent. If the tomb owners are Christians (especially fundamentalists), they may prefer to not use traditional motifs considered to belong to the old 'pagan' religion. They would prefer to place a wooden cross near the entrance of the *liang pa'* instead, although on several occasions we saw decorated tombs that integrated Christian crosses into traditional compositions, which shows that the two repertoires and traditions are not always considered as contradictory and mutually exclusive.

Traditional motifs on tomb doors

The motifs and techniques of tomb door decorations are directly borrowed from house wood carving tradition in Tana Toraja (see Nooy-Palm 1979, 238–240; Waterson 1988). They are produced by the same artisans, using the same execution techniques. This means that the motifs and decorative themes are not specific to the funerary sphere, their signification is not primarily associated with death. These motifs are associated with the domestic sphere and nobility, and with ideas of fertility, wealth and prosperity (Waterson 1988). Motifs are drawn from the natural world, in particular animals and plants. The repertoire of motifs decorating noble houses (*tongkonan*) and rice granaries (*alang*) is extremely rich with dozens of motifs and variants (Kadang 1960; Sande 1991). From this extensive repertoire, tomb decoration only borrows a small selection of motifs, which may be considered as the most important and most valued ones in Toraja iconography.

Here we will only discuss the main motifs appearing on tomb shutters. Decorated doors always have a single, large motif centrally placed on the door shutter. Smaller, secondary motifs are also added to create an elaborate composition around the main motif, on the shutter itself and on the wooden frame holding the shutter (Figs 3.5; 3.6; 3.8).

There are four main motifs, each with a specific name (Pls 5, 6; Fig. 3.8):

- *Pa'tedong*: The name of this motif means 'carved buffalo [head]', from *tedong* ('buffalo') and *pa'* ('to cut/carve', as in *liang pa'* 'cut tomb'). This is by far the most common motif on tomb doors. In our survey, it was found on 41 out of 60 old *liang pa'* (68%) and 199 out of 314 recent *liang pa'* (63%), accounting for the primary motif present on 240 out of 374 tomb doors overall (64%). Buffalo is the most important and valued animal in Toraja society. It is used mainly as a symbolic capital for exchanges and ceremonial sacrifices. *Pa'tedong* motifs bear the animal's symbolism of wealth, nobility and force and courage. They are often represented on door and window shutters of noble houses (*tongkonan*), rice granaries (*alang*) and rock tombs as apotropaic motifs to ward off evil influences from entering the buildings (Robin 2017).
- *Pa'barre allo*: The 'sunburst' motif or sun disc with rays (from *barre*, 'ray', and *allo*, 'sun'). This is the second most common motif. In the Sesean districts, it is found on three out of 60 old *liang pa'* (5%), 27 out of 314 recent *liang pa'* (9%), accounting for the primary motif present on 30 out of 374 tomb doors overall (8%). This heavenly motif has a special location on domestic buildings, appearing at the very top of the façade of the *tongkonan* house and *alang* rice barns (Waterson 1988, 54). According to Sande (1991) it 'symbolizes the grandeur and nobility of the Toraja people'.[2]
- *Pa'barana'*: Also called *Pa'barana' rapa'*, this motif represents a close formation of leaves of the banyan tree (*Ficus religiosa*), called *barana'* in Toraja language (Nooy Palm 1979, 243). This motif is not found on old *liang pa'* as a main door motif, but it occasionally appears on a smaller scale on the jambs or panel framing the door shutter (Fig. 3.5). On recent *liang pa'*, it is found as a main motif on 18 doors (6%) but is very often included as secondary ornamentation on the side and frame of tomb doors on which other designs are used as primary motifs.
- *Pa'kapu' baka*: From *kapu* ('knot') and *baka* ('basket'). This motif represents the tight bindings that secure the lid of the *baka* baskets. The latter are used traditionally inside houses to store special heirlooms, such as textiles, beads, gold ornaments or kris daggers (Waterson 1988, 47; Sande 1991). The motif symbolises the wealth of the *tongkonan* as a social group (Nooy Palm 1979, 239). We noted this motif on the door of seven recent *liang pa'* (2%). One motif on the door of an old *liang pa'* in Parinding may be a version of this motif (Fig. 3.6 left). We noted *pa'kapu' baka* carved as secondary motifs on the jambs of old tombs (Fig. 3.6 right).

3. The anatomy and decoration of liang pa' 67

Figure 3.8: Entrances of recent liang pa' rock-cut tombs with wood-carved door shutters. Top left: pa'tedong buffalo head motif (Buntu Lobo boulder 18, Sesean); top right: pa'barre allo sun motif (Bori' boulder 21, Sesean); bottom left: pa'kapu' baka basket-lid motif (Buntu Lobo 25, Sesean); bottom right: pa'barana' leaf motif (partly covered by a sun hat hanged on the door, Parinding boulder 7, Sesean) (photos: G. Robin).

Another traditional motif on the door of *liang pa'* is the representation of a human body in bas-relief, standing with the hands opened flat (Crystal 1985, fig. 160a). These can be regarded as 'fixed versions' of the *tau-tau* human effigies of the dead which were traditionally placed in front of *liang pa'* of nobles of the highest rank. The *tau-tau* wooden effigies (discussed in Chapter 5) are represented standing with the hands

Figure 3.9: Left: undecorated tomb door (Buntu Lobo boulder 12, Sesean); right: Christian cross painted on a tomb door (Batutumonga boulder 9, Sesean Suloara') (photos: G. Robin).

opened up in front of them in order to remind the living to come and bring them offerings. In only one instance, a recently hewn tomb in Lempo (boulder 10), did we see such human representation on the door of a tomb. However, we did see several old wooden tomb doors in local antique shops and private collections (Fig. 3.7). This particular motif is not found on domestic buildings, as it is a representation of specific deceased individuals and is exclusively associated with cemetery places.

Nowadays, Christian crosses are represented on the doors of *liang pa'*, sometimes as the only motif (Fig. 3.9) but more often as part of complex compositions that include traditional wood carving motifs. This particular type of decoration reflects a recent development which is part of broader changes in religious beliefs and funerary ritual practices in Tana Toraja (Waterson 1993).

Rock carvings

Decoration is normally limited to the wooden door of *liang pa'* but a few recently cut tombs display carved and sculpted motifs on the rocky part immediately around their recessed entrance. In our survey we recorded 21 tombs with rock carvings (Table 3.4). Rock decorations (and stone shutters) are not produced by the wood carvers who

3. The anatomy and decoration of liang pa'

Figure 3.10: Rock-cut tombs with stone-made door shutters. Top: Buntu Lobo boulder 20 (Sesean); bottom left: Buntu Lobo boulder 19 (Sesean); bottom right: Salu Liang (Malimbong Balepe') (photos: G. Robin).

produce tomb doors, but by stone workers who specialise in hewing *liang pa'*, and quarrying and carving menhirs (*simbuang batu*) and stone pillars that support the central front post of *tongkonan* houses (*tulak somba*).

The main type of rock-carved motif on tombs is a sculpted buffalo head and horns. In all the cases we encountered (× 12), the motif was positioned immediately underneath the square opening of the entrance recess (Figs 3.11–3.13). This motif can be considered as a stony version of the *kabongo'*, a wooden buffalo head sculpture fitted out with real horns, which appears on the façade of *tongkonan* origin houses. The *kabongo'* is not just a decorative ornament, it has a specific meaning in Toraja culture: it commemorates particularly expensive buffalo sacrifices held by members of the house at a funeral ceremony, and is therefore not displayed on all *tongkonan* (Waterson 1988, 49; Robin 2017). Buffalo head reliefs carved on *liang pa'* tombs can also be called *kabongo'*, although their association with rock-cut tombs is a recent innovation, not a traditional element of the Toraja culture. As a consequence, the practice of displaying a *kabongo'* on a tomb is not strictly related to the holding of particular buffalo sacrifices (as it is for house *kabongo'*). In practice, according to local stone workers, anybody who can afford it can have a *kabongo'* sculpted on their tomb. It is a matter of wealth rather than ceremonial achievement and history. However, according to one of our informants, it is unlikely that a family would pay for a *kabongo'* to be carved on their tomb if their own *tongkonan* does not have one; it would seem awkward or pretentious.

Another type of rock relief is a depiction of the individual or couple that commissioned the tomb (Figs 3.14 and 3.15). These can be considered as modern versions of the traditional free-standing *tau-tau* effigies, which are placed in front of the tomb to watch the living. The *tau-tau* wooden effigies are only produced and permanently displayed in front of a tomb if the most expensive rituals (involving at least 36 buffalo sacrifices) are held at the person's funeral ceremony (Waterson 2009, 456–457). Modern *tau-tau* sculpted on the rock face of tombs do not have the same signification, and here again they only represent an extra cost for the tomb's commissioner(s). During our survey we recorded five sculpted *tau-tau*: three depicting couples, and two depicting a single individual (Table 3.4).

Other rock carvings include low-relief depictions of traditional motifs (*pa'barana'*, *pa'tedong*, *pa'bulu londong*: Figs 3.13 and 3.14), as well as Christian crosses (Fig. 3.10). Unique architectural sculptures can be seen on boulders 10 and 21 in Bori', where tomb entrances are flanked by fully sculpted imitations of *sembang*, which are the upwardly pointing ends of longitudinal beams on *tongkonan* (Kis-Jovak et al. 1988, 124) (Fig. 3.16).

Rock carving at *liang pa'* sites is a recent, marginal phenomenon. According to our survey, it is never associated with the older *liang pa'*, and is only associated with 21 out of 314 recent tombs (7%). However, this practice may develop in future years. Indeed, rock carvings can be regarded as an additional way to increase the visual impact of family tombs. They also represent a significant expenditure: for instance, a *kabongo'* sculpted under the entrance of a *liang pa'* can cost 10 million Rupiah (c. 580

Figure 3.11: Buffalo-head reliefs (kabongo') sculpted into the rock under the entrances of rock-cut tombs. Top: Lempo boulder 8 (Sesean Suloara'); bottom left: cemetery of Lo'ko' Mata in Tonga Riu (Sesean Suloara'); bottom right: cemetery of Sele in Suloara' (Sesean Suloara') (photos: G. Robin).

Figure 3.12: Buffalo-head reliefs (kabongo') sculpted into the rock under the entrances of rock-cut tombs. Left: Bori' boulder 26 (Sesean); right: Parinding boulder 5 (Sesean) (photos: G. Robin).

GBP in 2017). This adds to the already important costs of cutting a tomb (*c.* 60 million Rupiah, *c.* 3490 GBP). As a highly visible and expensive type of ornamentation, rock carvings are explicit displays of wealth. They represent another sign of the inflation in ceremonial expenditures that have marked social life in the past few decades in Tana Toraja.

Paint and paintings
Paint is often used in recent tombs but seems totally absent from older ones. Four main colours are used (black, red, white, yellow), which are applied to enhance wood and rock carvings representing traditional motifs (Pls 5, 6). Paint is rarely found on its own, *i.e.*, not in association with carvings. When it does, it is used to colour over plain wooden doors, or to paint Christian crosses (Fig. 3.9). This indicates that carving is the only technique used to create traditional Toraja motifs on tombs. For these motifs, paint is only used as a secondary enhancement, not as a primary technique.

Regional patterns in tomb decoration
Tomb carvings use motifs from a single and widely shared cultural repertoire, which combines traditional Toraja motifs and Christian religious symbols (to a lesser extent). There is, however, a certain level of diversity from one tomb to another, and even from one community area to another. Does variability in tomb decoration reflect cultural

3. The anatomy and decoration of liang pa'

Figure 3.13: A richly decorated tomb on boulder 7 at Lempo (Sesean Suloara'), with rock-carved kabongo' *(buffalo-head sculpture),* pa'barana' *(leaves),* pa'bulu londong *(feathers) motifs, surrounding a wooden door adorned with the* pa'barre allo *(sun) motif (photo: G. Robin).*

preferences and local identities? It is difficult to be conclusive, but it is interesting to note a few patterns within the relatively small sample of our 2017 survey in area A. Specific motifs are sometimes selected or ignored locally. For instance, the *pa'barre allo* (sunburst) motif is particularly frequent in Bori', Buntu Lobo and Parinding, three

Table 3.4. List of tombs with rock-carved motifs in the Sesean districts (study area A).

Location	Tomb ID	Kabongo'	Tau-tau	Other
Bori'	BO10.1	–	–	Sculpted *sembang* house beams
Bori'	BO21.10	–	–	Sculpted *sembang* house beams
Bori'	BO26.3	1	–	–
Buntu Lobo	BL07.3	1	–	Carved *pa'barana', pa'bulu londong*
Buntu Lobo	BL11.1	–	–	Sculpted chair
Buntu Lobo	BL20.3	–	–	Sculpted Christian crosses
Buntu Lobo	BL20.4	–	–	Sculpted Christian crosses
Buntu Lobo	BL21.1	–	1	–
Buntu Lobo	BL31.2	1	–	–
Deri	DE02.5	1	–	Carved *pa'barana', pa'bulu londong*
Deri	DE02.6	1	–	–
Deri	DE13.2	–	1	–
Lempo	LE08.2	1	–	Sculpted Christian crosses
Lempo	LE10.1	–	1	–
Lempo	LE14.1	1	1	–
Lempo	LE41.1	–	1	Carved *pa'barana', pa'tedong, pa'bulu londong*
Parinding	PA05.4	1	–	–
Parinding	PA05.5	1	–	–
Parinding	PA16.1	1	–	–
Tonga Riu	TR01.43	1	–	–
Suloara'	SU02.10	1	–	–
Total	21	12	5	9

groups that are geographically close to each other (for location names, see map in Fig. 2.17). By contrast, this motif is totally (and surprisingly) absent in Batutumonga and Suloara'. The *pa'barana'* (spiral) motif is more frequent in Buntu Lobo and Tonga Riu, and is absent in Batutumonga and Deri. The *pa'kapu' baka* (basket lid) motif is extremely rare or absent in most locations (1 or 0 occurrence), except in Tonga Riu (4 occurrences). Similarly, the Christian cross is very rare everywhere, except in Batutumonga: does this reflect the presence of a particularly devout Christian community there?

Plain doors are very common in certain locations (Suloara': 35% of recent *liang pa'*; Buntu Lobo: 26% of all tombs; Lempo: 21% of all tombs; Batutumonga: 19% of recent *liang pa'*). By contrast they are very rare in Bori' (1%). Finally, and more understandably, rock decorations are particularly common in Sesean's mountainous areas that are reputed all over Tana Toraja for stone carvers, especially Buntu Lobo and Lempo.

Figure 3.14: Rock-carved pa'barana' *(leaves) and* pa'bulu londong *(feathers) motifs surrounding the entrance of a tomb on Lempo boulder 41 (Sesean Suloara'). Note the human face and upper torso (likely portraying the tomb's sponsor), reminiscent of the* tau-tau *tradition (photo: G. Robin).*

Although specialised stone workers from these areas typically work outside their village and can produce rock carvings in distant cemeteries (see Chapter 4).

Conclusion

The main conclusion that can be drawn from this detailed description is that *Liang pa'* rock-cut tombs are highly standardised. There is very little diversity in their design and decoration. Their architecture is based on the same general model, that of a rectangular box just deep enough to receive bodies in supine position. Recent tombs are almost always created as a cube with the exact same dimensions (2 × 2 × 2 m). Only a few tombs include minor alterations to this model, with the creation of elevated recesses for single coffins, or larger chambers. Decorations too are very standard. They are based on only four motifs taken from the broader house-carving repertoire (why are no more motifs from that repertoire used on tombs?). Very few carved design novelties have been added in recent decades: stone carvings of *kabongo'* and *tau-tau*, and Christian crosses.

Figure 3.15: Rock-carving surrounding the entrance of a tomb in boulder 14 in Lempo (Sesean Suloara'). The two human figures on each side of the entrance represent the couple who sponsored the cutting of the tomb (photo: G. Robin).

This standardised character is intriguing and contrasts with that of *patane* house-tombs, whose recent concrete-built versions show a large degree of variability (if not eccentricity in many cases) in their size, architecture and ornamentation. In comparison, *liang pa'* appear far more conservative and traditional. We also noted that *liang pa'* are still considered as the most prestigious type of tomb today. The main impact of recent economic development in Tana Toraja has not been so much on the aspect of *liang pa'* than on their number, with an exponential increase both within traditional status groups (noble families) and beyond (individuals with newly acquired wealth).

The highly standardised character of *liang pa'* suggests two points. First, their funerary function is rather simple: they do not require elaborations or multiple social or ritual apparatuses. They are primarily containers for mummified human bodies. As we will see in the next chapter, all the complexity and elaboration of funerary rituals take place outside the tombs, not inside them. Second, unlike other forms of tombs, such as *patane*, *liang pa'* act as very codified social markers that reproduce and maintain long-established, conservative conventions. In recent decades, a large number of families and individuals have adopted this marker and seem to have resisted deviating from the traditional model in order to be more efficiently and explicitly associated with the prestige of the old nobility.

3. *The anatomy and decoration of* liang pa' 77

Figure 3.16: Rock-carved sembang *(house-beams) on each side of a tomb entrance in Bori' boulder 10 (Sesean). This recent (and rare) type of sculpted decoration emphasises the notion of tombs as houses of the dead (*tongkonan tangmerambu, *'houses with no smoke') (photo: G. Robin).*

Notes
1. Similarly, Waterson reports that despite the loosening of social restrictions regarding traditional levels of funeral ceremonies, many people said they would be ashamed to hold a ceremony to which they were not entitled as lower nobles, commoners or descendants of slaves (Waterson 2009, 383).
2. According to Kruyt (1924, 164), *pa'barre allo* were also used in the Mamasa region to decorate wooden house-tombs specifically associated with a noble woman. In our experience, such a gender association with the motif does not prevail in the Sa'dan region.

Chapter 4

Creating a *liang pa'*: cutting process and rituals

This chapter discusses the process of creating a *liang pa'* rock-cut tomb. It presents the context and key actors involved, from the sponsors commissioning the new tombs to the specialised stone workers creating them. The chapter describes the different steps involved in this complex process, from reserving a location on a rock face, to negotiating the costs, organising and carrying out the cutting work and using the extracted stone material for other purposes. We address not only the technical aspects of cutting a new tomb, but also its economic, social and ritual implications.

This topic has been only partially and superficially addressed in the anthropological literature (Kruyt 1924, 164; Keers 1939, 207; Nooy-Palm 1979, 260; Waterson 1995, 207). One of the priorities of our 2017 fieldwork was to tackle this data gap and to clarify five primary issues:

- *Tomb creation*: when is a new rock-cut tomb needed and created? Does that happen only when a new *tongkonan* is created? Or can a family decide to build a second, new tomb if the old one is full? Are there any specific rituals associated with the creation of a rock-cut tomb?
- *Location choice*: how does the local geology (limestone *vs* basalt) influence the location, dimensions and morphology of rock-cut tombs? Within a village community area, or at the scale of a cemetery site, to what extent is the quality of the rock taken into account in these choices? Can one excavate a tomb anywhere in one's community cemetery, or is this the object of prior negotiation? Can rock faces within the cemetery be owned or reserved by individuals prior to the creation of a tomb?
- *Wealth and tomb size*: how does variability in sponsors' wealth impact the size, architecture and decoration of newly created tombs across Tana Toraja? Can tombs be altered during their lifespan, for instance be recut to be expanded to accommodate more burial depositions?

- *Stone workers*: who are the specialised workers cutting the tombs? Are they part of the village community (does each village have a 'tomb maker'?), or are they from elsewhere? How are they paid and how much does it cost overall to create a rock-cut tomb? Are stone workers also responsible for the wooden doors that close the tombs, or are these made by another category of artisans?
- *Cutting work*: what tools or technologies are used to cut the rock? How many workers are involved in the work? How is the work organised and what are the main steps? Are there any rituals involved in the process?

The majority of the cemetery sites we visited included old and recent rock-cut tombs. Some of them were in the process of being cut (67 tombs in total[1], *i.e.*, roughly 10% of all the tombs we surveyed: Fig. 4.1). By combining visits at different sites, we were able to document the different stages of progression in the cutting work of a traditional *liang pa'*. We also carried out semi-structured interviews with stone workers at tomb cutting sites in cemeteries in Bori', Buntu Lobo, Lemo, Lempo and Tonga Riu (Lo'ko' Mata). Interviews were also conducted with tomb owners and sponsors and ritual specialists (*to minaa*).

It is difficult to know to what extent the entire process and methods used to create rock-cut tombs have changed throughout the four centuries of the *liang pa'* tradition. We were often told that the emergence of *liang pa'* in the 17th century corresponded to the introduction of metal tools in Tana Toraja, which made it possible to cut into the rock (Waterson 1988, 37). Metal tools are still used today to hew out tombs, and it is likely that methods have not changed drastically over the two main generations of *liang pa'* (Figs 4.1 and 4.2). The larger size and added stone sculptures characterising the more recent generation may correspond to some improvement in technologies but are more likely related to the significant economic growth in Tana Toraja from the late 1960s onwards (Waterson 2009, 117–118). This growth has resulted in a rise in available finances to support increasing ceremonial expenditures and status display between competing house kinship groups (see Waterson 1988 for a similar observation on the evolution of *tongkonan* houses' dimensions, roof style and decorations). Below, we describe the different steps involved in the creation of a tomb, from identifying its location in a rock face, to cutting it out and consecrating it. The information collected from stone workers is more relevant to the more recent generation of *liang pa'*, but several aspects, such as the ritual involved in the cutting process, are rooted in older traditional practices.

When and where are new *liang pa'* created?

Each *liang pa'* rock-cut tomb in Tana Toraja is associated to a specific *tongkonan* house (Waterson 1995), which is located in the vicinity of the cemetery. Practically, one *liang pa'* cannot be associated to two different *tongkonan*, but one *tongkonan* can have several *liang pa'*, for example if an old one is full. As far as we could gather (see also Jeunesse and Denaire 2018, 100), there are two main reasons for creating a new *liang pa'* tomb: either the old tomb of a *tongkonan* is full; or a new family branch is created and a new

Figure 4.1: Liang pa' *rock-cut tomb in the process of being cut at Lemo (Makale Utara) in June 2017 (photos: G. Robin).*

tongkonan is founded, which requires the creation of its own *liang pa'*. However, with recent economic developments, wealthy individuals may also decide to create a new tomb solely for their own nuclear family for the sake of comfort rather than necessity.

4. Creating a liang pa': cutting process and rituals

In any case, the creation of a new tomb is sponsored and supervised by a senior member of the family who is responsible for paying the stone workers carving out the tomb. Roxanna Waterson highlighted interesting parallels between founding a *tongkonan* and founding a *liang pa'* tomb:

> Just as the founders of a *tongkonan* are remembered, so too is the maker of a *liang*. A house founder may be referred to as *to mangraruk* or *to umpabendan*, 'the one who erected', and the commissioner of a *liang* as *to pa'pa'na*, 'the one who pierced'. Quite often a *liang* is known by the name of its maker; *liangna Dondan*, or 'Dondan's liang', for example, refers not to any particularly famous ancestor buried in this grave but to its original maker. (Waterson 1995, 207)

Figure 4.2: Stone workers cutting a liang pa' rock-cut tomb around 1935 at an unknown location in Tana Toraja (photo: unknown author, public domain, Leiden University Libraries).

Once the decision to make a new tomb is made, one needs to select an appropriate location. The main factors here are geomorphology and land ownership. In mountainous areas north of Rantepao, where small volcanic rock boulders are scattered across rice fields, landowners can simply select a boulder on their property (as long as it is sufficiently distant from houses). This practice is less frequent in the limestone districts south of Rantepao, where rock faces are comparatively smaller in number and large cliffs are used as communal cemeteries by several *tongkonan* groups. Particularly large boulders or outcrops located on private lands may be shared with other *tongkonan* with which the owners have good relationships. For instance, the cemetery of Sele in Suloara' (35 tombs belonging to at least eight different *tongkonan*) is on the land of one particular kinship group (*tongkonan* Rante Bulaan). In the past, other *tongkonan* groups could use the cemetery if they had good relations with the kinship group that owned the land, with no need to pay them any fees to create a *liang pa'*. Nowadays, perhaps because available rock surfaces are becoming rare, only members of *tongkonan* Rante Bulaan can cut new tombs at Sele.

Other cemeteries seem to have retained their original 'public' character, being owned not by a specific *tongkonan* but by a community of several *tongkonan*. For instance, the cemetery of Batu Lappa' in Buri' has 14 tombs that belong to at least seven different *tongkonan*. In the past, any *tongkonan* in the vicinity could come freely and create their *liang pa'* where space was available on the rock face. However, rights have become more restricted recently, and only *tongkonan* who already have a tomb in Batu Lappa' are permitted to cut a new *liang pa'* (see Chapters 6 and 7 for further information).

The specific location of the new tomb within a rock face is also a matter of careful consideration and negotiations. The sponsor of the tomb will ask the advice of an experienced stone cutter who has the knowledge to identify a good location for cutting a *liang pa'*, based on the quality of the rock. Traditionally, if the selected rock face or boulder had never been cut for a tomb before, the sponsor would need to ask permission from the *deata*, the deities of the natural world. To do so, a *kamboja* plant is transplanted in the ground in front of the proposed *liang pa'* location. If the plant does not sprout flowers or branches after three days, then it is considered to be a message from the *deata* that the project should not be permitted to proceed and that another location must be sought.

It is possible for a sponsor to reserve a spot on a rock face in advance of the cutting work. This process is called *ma'suri* ('to reserve') and is made by a professional stone cutter (not necessarily the same one that will be hired to cut out the tomb) for a cost of *c.* 200,000–300,000 IDR (*c.* 12–17 GBP or 15–22 USD[2]). This consists of creating a mark that corresponds to the outline of the future entrance opening of the tomb. Different techniques have been observed: carving out a shallow rectangular surface, carving two vertical parallel lines, or painting a rectangle (Fig. 4.3). Such actions typically serve to

Figure 4.3: Carved and painted rectangular marks used to reserve locations for future rock-cut tombs on cemetery rock faces (photos: G. Robin).

'reserve' a location for a long, undetermined time, often several years. In some cases, reserved spots never end up getting used at all. We noted one in the old cemetery of Pana' in Suloara': this cemetery was abandoned several generations ago and no one will use the remaining unused reserved spot there in the future. The tomb sponsor may not have been able to assemble the finances to build the *liang pa'* there or may have decided to relocate it to the newer cemetery 150 m away (Sele).

The stone workers: costs, workspaces, tools and roles

Liang pa' are cut by stone workers specialising in tomb carving and standing stone (*simbuang batu*) quarrying. In the Sa'dan region, such specialists are based in Sesean's mountains, in particular in the *lembang* of Bori', Lempo, Tonga Riu and Buntu Lobo, as noted in earlier ethnographic works (Keers 1939, 207; Nooy-Palm 1979, 260). Costs to create a rock-cut tomb are negotiated months or years in advance of the project. Traditionally, and until recently, stone workers were paid in buffaloes. In the early 1900s, the wage was one buffalo per stone worker, so one tomb would cost two or three animals in accordance with the size of the crew contracted to cut the stone (Kruyt 1924, 162). More recently, up to seven buffaloes were paid for one tomb, which corresponded to a cash value of 42–59 million IDR in 2000 (Waterson 1995, 207; 2009, 409; Duli 2018, 44). According to our own information collected in 2017 at ten worksites, stone workers are only paid in cash nowadays, with foodstuffs provided to the workers as an additional expense (rice to cook and eat while working). A typical tomb costs 60 million IDR (3480 GBP or 4480 USD), although this figure can vary according to the dimensions of the chamber, the hardness of the rock, and the distance between the worksite and the home of the workers (Table 4.1).

A minority of tombs have additional features which involve extra expenses. For instance, some sponsors require an upper recess or bench to be cut into the back wall or a side wall of the chamber. As discussed earlier, this extra space is reserved for their own body exclusively, so it does not get crushed by the stacked wrapped bodies or coffins of other family members to be deposited progressively in the main chamber by future generations. We estimate the extra cost for this space to be 20–40 million IDR. A buffalo head sculpted underneath a tomb entrance in Bori' cost 10 million IDR and took 3 months to be completed (see Fig. 3.12). The wooden doors that close *liang pa'* tombs are made by specialised wood carvers and represent a separate expense.

When tomb cutting work commences, the crew of stone workers creates temporary living quarters in front of the rock face (Fig. 4.4). This includes a shelter made of bamboo and leaves for their workshop (a small hearth to sharpen metal picks), kitchen and sleeping area, a garden to grow chillies and vegetables to be cooked and eaten with rice and an open-air bathroom area enclosed by vertical tarpaulins. If distant from home, workers sleep overnight in the shelter.

Cutting a *liang pa'* requires an experienced know-how that is normally transmitted from father to son, or another male member of the same family. Apprentices begin the craft at a very young age and are paid in rice; they become full-fledged stone cutters at the age of *c.* 20. Workers use iron pick tools with a hammer (indirect percussion) (Fig. 4.1). Picks of various lengths are used for different tasks (Fig. 4.5). Shorter picks are used for delicate and accurate work, such as cutting out the entrance doorway. Longer picks are used for rougher stone extraction work, for instance when hollowing out the chamber space. When cutting one tomb, workers are usually organised in a crew of three with specific roles and tasks taken in rotation: one is cutting the stone, the second is hauling out stone blocks that were cut from the interior of the tomb and the third worker is resting, cooking or sharpening tools (Fig. 4.6). Work payment is shared equally among the crew, and there does not seem to be any specific hierarchy, although senior stone workers can sometimes have more strategic roles, such as liaising with tomb sponsors and sometimes sub-contracting cutting work to other, more junior, crews.

Tomb cutting work can happen throughout the year; there are no set cutting seasons or particular religious prohibitions imposing break periods. The duration of the cutting work varies from 2 to 8 months nowadays, depending on the volume of the chamber (Table 4.1) and the size of the crew (2–4 members). Some crews may also work on several tomb projects in parallel, which affects the completion time of

Table 4.1. Data collected at ten rock-cut tombs in process of being cut in June 2017 (NR = not recorded).

Locations	Cemetery	Dimensions (d/w/h) and volume	Work duration (months)	Costs	Carver's home
Lempo	Boulder 17	2.30 × 3.00 × 2.50 m (17.25 m^3)	8	95 million IDR	Lempo
Lempo	Boulder 17	2.20 × 3.00 × 2.00 m (13.20 m^3)	5.5	85 million IDR	Lempo
Lempo	Boulder 17	2.00 × 2.00 × 2.00 m (8.00 m^3)	2	55 million IDR	Lempo
Lempo	Boulder 18	2.20 × 3.00 × 2.20 m (14.52 m^3)	NR	60 million IDR	NR
Lempo	Boulder 18	2.20 × 2.00 × 2.00 m (8.80 m^3)	NR	40 million IDR	NR
Lempo	Boulder 18	2.20 × 3.00 × 2.00 m (13.20 m^3)	NR	60 million IDR	NR
Buntu Lobo	Boulder 1	2.20 × 3.00 × 2.00 m (13.20 m^3)	5	100 million IDR	Buntu Lobo
Buntu Lobo	Boulder 20	2.20 × 3.00 × 2.00 m (13.20 m^3)	6	70 million IDR + 300 kg rice	Buntu Lobo
Tonga Riu	Lo'ko' Mata	NR	NR	50 million IDR	Tonga Riu
Tonga Riu	Lo'ko' Mata	NR	NR	50 million IDR	Tonga Riu
Lemo	Cliff B	2.00 × 4.00 × 2.00 m (16.00 m^3) + bench	3–4	110 million IDR + 200 kg rice	Riu (Tonga Riu)

4. Creating a liang pa': cutting process and rituals

Figure 4.4: A liang pa' worksite at Suloara' boulder 10 (Sesean Suloara') (photos: G. Robin).

individual projects. *Liang pa'* from the older generation are smaller (Chapter 3), but the time to complete them might have been similar, if not longer, according to various authors: from over 1 year (Wilcox 1949, 84) to about 2 years (Brisbois and Douvier 1980, 116) and up to 3 years if the tomb was located high up in a cliff face (Koubi 1982, 196).

Cutting out the tomb: a technical and ritual *chaîne-opératoire*

Figure 4.5: Metal picks and hammer used by stone workers to hew out rock-cut tombs (photo: G. Robin).

Toraja rock-cut tombs are relatively simple underground structures, with an entrance opening and a single rectangular chamber with plain flat walls and ceiling. Nevertheless, the actual cutting work to create them is not a single, straightforward process. This process involves a sequence of distinct, well-defined steps, each having a specific name

Figure 4.6: Stone workers work in rotation: while one cuts the stone inside the tomb, the second hauls out stone blocks from the interior of the tomb and disposes them outside, as here in Lemo (photos: G. Robin).

and requiring a dedicated ritual. Such a complex *chaîne-opératoire* must be considered not only as a process of 'building', but also as a process of progressive negotiation between the carvers and the deities of the natural world (*deata*: Waterson 2009, 141). Cutting a tomb is an intrusive action that alters the rock, opens up and penetrates into the mineral world and therefore creates a disturbance that can be taken as an offense if appropriate rituals are not carried out as a compensation. Indeed, if the deities are not happy with this intrusion and disorder, they can provoke a failure in the cutting process, such as a crack in the rock, or water infiltrating the rock in a way that will make the tomb unusable. To ensure its success, the cutting work is carefully divided into a series of technical steps, to which a specific ritual intended for the deities of nature is attached, and which involves the sacrifice of an animal. In general, the stone cutting crew slaughters the sacrificed animal, although the sponsoring family also attends and eats meat and rice (and provides the animals and rice for the rituals). The function of these sacrifices is to thank the deities for allowing humans to carry out the intrusive cutting work into the rock. They also enable the stone workers to be retributed with meat throughout the cutting work. This process culminates with the consecration of the tomb. Note that a similar sequence of rituals marks the phased construction of traditional *tongkonan* houses in Tana Toraja (Nooy-Palm 1979, 244–252; Koubi 1982, 195).[3]

The sequence of cutting work and rituals was explained to us by Pak Saipan, a senior stone worker we met in Bori' (Table 4.2 and Fig. 4.7). At the start of the process a dog

Table 4.2. Names of the main steps required for creating a liang pa' rock-cut tomb, with their associated stone cutting and ritual activities.

	Step and ritual name	Stone cutting activities	Ritual activities
1	Asking permission to nature	None	Branch of *kamboja*, dog sacrifice
2	*Ma'suri* ('to reserve')	Tomb entrance location inscribed on the rock face	None
3	Start of cutting work	Entrance doorway	Dog sacrifice
4	*Ma'siku* ('elbow')	Anterior corners of chamber	Chicken, pig or dog sacrifice
5	*Ma'parampo* ('to arrive at destination')	Back corners of chamber	Chicken, pig or dog sacrifice
6	*Ma'rinding* ('wall')	Completion of the side and back walls	Chicken, pig or dog sacrifice
7	*Ma'bubung* ('roof')	Completion of the ceiling	Pig sacrifice
8	*Ma'sali* ('floor')	Completion of the floor	Pig sacrifice
9	*Di tutu'i* ('to close with a door')	Wooden door fastened to entrance doorway	Dog sacrifice
10	*Massabu* ('to consecrate/inaugurate'), with *mangrara* ('to anoint with blood')	None	Pig sacrifice, chicken blood splattered inside tomb

is killed by the work crew just prior to commencing the stone cutting. Pak Saipan's crew had killed and eaten a dog just prior to beginning their work a few days before our visit and banana leaves were still lying around outside the entrance of the *liang pa'* that would have been used as plates for the dog meat eaten for the small feast. The first step of the sequence consists of cutting the entrance doorway, which is the most difficult and time-consuming part of the entire process. The next ritual is called *ma'siku* (from *siku*, 'elbow' in Torajan) for which a chicken or pig is killed and eaten. This ritual is held when the cutting crew establishes the front corners (near the door) of the *liang pa'* chamber, for which workers traditionally use their elbow to set the meeting walls at the right angle. In the past, women could not eat the meat served at a *ma'siku*, as it could lead to problems with a future pregnancy. This restriction does not apply to the other rituals performed in the process of cutting a *liang pa'*. When the crew subsequently establishes the corners in the rear of the chamber, they hold a ritual called *ma'parampo*, which, in that context, can be translated as 'to arrive where you are intending to arrive' (from *rampo*, 'to arrive'). The ritual involves the slaughter and eating of a chicken or a dog.

Ma'rinding ('to make the walls') is the ritual held once the wall surfaces between the four corners of the chamber are straightened and completed. It involves again the killing and eating of a chicken or a pig. The subsequent step is the straightening of the ceiling surface of the chamber: this is marked by a ritual called *ma'bubung* (from *bubung*, a roof ridge made of flattened and folded bamboos), which is also performed after the completion of the roof of a *tongkonan* house (van der Veen 1940, 77; Nooy-Palm 1979, 244; Waterson 1990, 127–129). This ritual requires a pig sacrifice. Then, the chamber floor surface is levelled and completed, which represents the last step in the sequence of stone cutting work required to create the tomb (Fig. 4.8). The dedicated ritual for this step involves the sacrifice of one or several pigs that are eaten in a small feast referred to as *ma'sali* (or *massali*, 'to lay out the floor': Nooy-Palm 1986, 66). The word *sali* means 'floor' in Torajan and refers either to the elevated floor that is laid beneath rice barns (*alang*) and which is often used as a social space to meet and drink coffee, or to the central room of the *tongkonan* (van der Veen 1940, 558–559; Nooy-Palm 1979, 252–254, 276). The final step in constructing a *liang pa'* tomb is the placement and fastening of the wooden door (Fig. 4.9). This is celebrated with a ritual called *di tutu'i* ('to close (with a door)') and which involves a dog sacrifice.

Consecration of the tomb

Toraja tombs can be completed several months or years before they are eventually used for the interment of the deceased. During this interval between tomb completion and interment, they are normally left open, which conveniently allowed us to observe and record several tomb interiors during our visit in 2017 (Fig. 4.10). At this stage, tombs are not considered sacred spaces, they are only *lo'ko'*, simple 'holes' in the rock. It is only after the consecration ceremony that they become proper *liang* ('tombs'), prohibited spaces that can only be open and visited on rare ceremonial occasions.

4. Creating a liang pa': cutting process and rituals

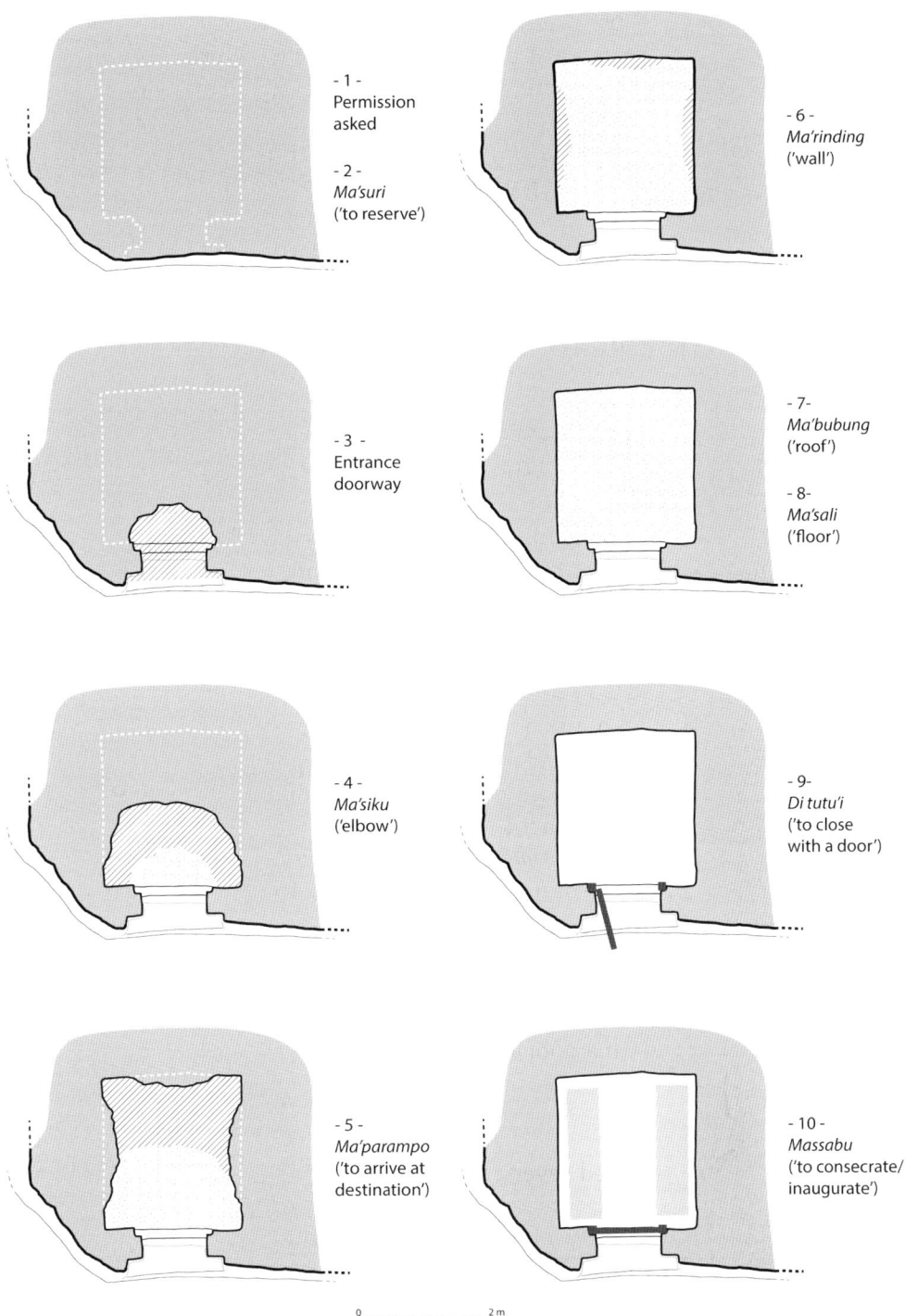

Figure 4.7: Sequence of cutting work for creating a liang pa' rock-cut tomb (image: G. Robin).

Figure 4.8: Cutting and flattening the floor of the chamber is the last part of the hewing process, as here in Bori' boulder 1 (Sesean) (photo: G. Robin).

The tomb consecration, *massabu* ('to consecrate, to inaugurate'),[4] is a ceremony at which a pig is killed and eaten to give thanks to nature for allowing a successful *liang pa'* construction. We observed this celebration in Lempo, which was attended by c. 20 people from the family of the tomb sponsor. The stone workers were not present on this occasion, but a female Christian minister joined for a short time and performed a sermon in front of the *liang pa'*. She led everyone in prayer in hopes that the good spirit in the *liang pa'* would help the dead ancestors and the living family. She gave thanks to the family for hosting the event and to the carvers for creating the tomb. She prayed for the good health of the family of the deceased as well. In the meantime, a small pig had been brought to the location of the tomb and subsequently slaughtered and butchered. The meat from the pig was cooked in bamboo cylinders (*pa'piong*) and served on banana leaves.

After the small feast, and as part of the whole *massabu* ceremony, a *mangrara* ('to anoint with blood') ritual was held to mark the actual act of consecrating the tomb. This ritual is also celebrated to consecrate newly built *tongkonan* houses (*mangrara banua*), although under a more elaborate form (Nooy-Palm 1979, 248–250; Waterson 2009, 195–200). At Lempo, a living rooster was brought to the *liang pa'* to be consecrated; a small cut was made on its neck, and the animal was promptly held inside the tomb chamber and shook firmly in order to splatter blood on the interior walls and threshold.

4. Creating a liang pa': cutting process and rituals

Figure 4.9: Installation of a wooden door on a rock-cut tomb in Lempo boulder 18 (Sesean Suloara') (photo: G. Robin).

The normal procedure is to cut, but not kill, the rooster: after the incision, the rooster we observed appeared largely unharmed and went about its business.

The *massabu* ceremony is typically held within a relatively short period of time before a body is to be deposited in the tomb. We witnessed the placement of the body in the tomb in Lempo nine days after the *massabu*, on the third day of the funeral feast. The placement of the wrapped body in the rock-cut chamber (Pl. 8) and the subsequent closure of the tomb door were carried out rather promptly, with no particular ceremonial protocol. This was in marked contrast to the slow sequence of dramatic rituals that had preceded (including the procession of the *saringan* palanquin from the house to the cemetery: Nooy-Palm 1979, 261).

The consecration of the tomb definitively marks the end of its construction process. Once a *massabu* ceremony has been performed, it is no longer possible to do any subsequent cutting work, such as an extension of the chamber or the addition of a carved buffalo head. The tomb becomes 'petrified' and can no longer be altered. This does not mean that tomb extensions never happen, but they are very rare and must be completed prior to consecration. In Bori', we observed a *liang pa'* that was originally completed 23 years prior to our 2017 visit but had remained vacant since then and had not yet been consecrated. Its sponsor decided to expand its depth by an additional 2 m and paid 50 million IDR for the cutting work (which was in progress during our visit).

Figure 4.10: A recently completed liang pa' *rock-cut tomb in Lempo (Sesean Suloara'). It will be left open until its consecration, which may take place many years after the end of the cutting work (photo: G. Robin).*

Cutting failures and abandonments

During our survey of *liang pa'* cemeteries, we observed several instances of tombs for which the cutting work had been interrupted halfway through and left incomplete for several years. Such uncompleted tombs are typically the result of two primary circumstances: 1) during the cutting, the workers determine that the selected rock is of poor structural quality or is affected by infiltrating water (this can be most common in karst contexts around Rantepao), causing the project to be abandoned indefinitely; 2) the family sponsoring the work runs short of resources during the course of the cutting work and is no longer able to pay the stone carvers, in which case the project can be suspended provisionally for years or decades, or completely abandoned.

We also observed vestiges of cutting failures. We were told that tomb carving work may be stopped as a result of an incident during the cutting process, typically a break in the rock originating from the cutting work itself (possibly facilitated by pre-existing weaknesses in the rock structure), causing a large part of the tomb to break off the rock face and fall to the ground, an example of which we saw in Bori'. Other rock-cut tombs that had been fractured into multiple pieces at Salu Liang and Batu Lappa' cemeteries were described to us as having been struck by lightning.

Unfortunately, we were not able to interview stone workers or sponsors that had experience with such cutting failures. Considering all the ritual precautions involved in creating a *liang pa'*, it would be interesting to know how these failures are interpreted (*e.g.*, as results of rituals unsatisfactorily conducted, or as manifestations of disagreements from the ancestors or nature deities), whether they are considered as pollution (potentially impacting the social reputation of the sponsor) and whether specific rituals are typically carried out subsequent to cutting failures.

Tomb worksites as stone quarries

Unlike most examples of stone monument construction, creating a rock-cut tomb is a *subtractive* process rather than an *additive* one. Undesirable stone material is detached and removed from the rock in order to create the tomb interior space. What is the fate of this unwanted material? In several instances, the extracted stone material was used at the front of the tomb by the stone workers in order to create a platform facilitating access to the tomb entrance area (Figs 4.1 and 4.10). However, the majority of the material is simply disregarded by the workers and can be taken away by members of local communities to be used at villages or in the fields. People do not need to pay or to be related to the family sponsoring the stone cutting in order to take this material. The extracted material can serve different purposes: polygonal blocks can be used to reinforce terraces or retaining walls in rice paddies or banks of streams; small stone chips are typically collected to be spread over mud in hamlet courtyards or access paths (Fig. 4.11). This practice of collecting stone from tomb worksites is tolerated while the tomb building is underway. However, once tombs are consecrated, removing the extracted stone material is no longer permitted, even if the stone is simply lying in front of the tomb without serving any purposes. The domain of the consecration, therefore, is extended beyond the tomb itself to its extracted fragments.

Conclusion

The process of creating *liang pa'* tombs is more complex than one could expect, considering the simplicity of the architecture of the tombs. It involves a series of both technical and ritual activities that are tightly intertwined. The enterprise is not only one involving stone cutting, but also a permanent negotiation with the deities of nature. As a consequence, workers are not only responsible for cutting the stone but also for conducting small rituals.

Moreover, tomb cutting has a noteworthy economic dimension, not only for the parties directly concerned (sponsors and stone workers negotiating cutting costs) but also for local communities: for them, tomb worksites are also small stone quarries that can be used freely to obtain convenient stone blocks and chips to improve (rather than to really build) areas of their domestic structures and rice fields.

This study also highlights the long-noted conceptual connections between houses of the living (*tongkonan*) and tombs as houses of the dead (*tongkonan tangmerambu*).

Figure 4.11: Top: woman collecting stone chips at a liang pa' *worksite in Bori' (Sesean); bottom: field retaining wall in Suloara' (Sesean Suloara'), built from stone blocks that were extracted during the cutting of the tomb visible in the background (photos: G. Robin).*

This close connection is not materialised in the formal aspect of the tomb, apart from the decoration of the wooden door (*pa'tedong* motifs) and rare sculpted buffalo heads (*kabongo'*), which replicate *tongkonan* decorations (Waterson 1988). In contrast, the simplicity of the stone chamber and of the tomb exterior differs markedly from the conspicuous architecture of the kinship houses with which they are paired (see Fig. 1.9). The connection between the tombs and the houses is nevertheless very explicit in many ways, and this chapter shows how this is achieved through ritual terminology marking the cutting steps, which directly reference house building rituals.

Finally, the information we collected highlights the particular social status and value of stone in Tana Toraja. According to traditional religion (*aluk to dolo*), mythical ancestors of Toraja people originated from the mineral world. Being buried in stone is like returning to these origins. Stone (*batu*), as opposed to other materials, such as wood (*kayu*), is associated with the origins, the ancestors, permanency and immortality (Waterson 2009, 130–131). Stone is used for standing stones (*simbuang batu*) that commemorate large funeral feasts. It is also the preferred material for burials, namely rock-cut tombs. *Liang pa'* are more valued, for example, than modern concrete house-tombs (*patane*) that have become popular in the last few decades as a more economical and convenient alternative. Interestingly, the status of rock material associated with tombs is transformed throughout the process of creating and completing the tomb. After its consecration, the tomb itself moves from the status of a simple hole (*lo'ko'*) to that of a sacred burial space (*liang*), and the stone extracted during the process moves from the status of free refuse material to that of untouchable extensions of the sacred tomb.

Creating a rock-cut tomb in Tana Toraja, therefore, is not just a technical enterprise: it is also an expression of traditional beliefs, of the ritual relationship between the Toraja and their landscape and the social value of stone as a material associated with death and status.

Notes

1. *Lembang* where we observed tombs in progress of creation: Bori', Buntu Lobo, Deri, Lemo, Lempo, Parinding, Tonga Riu, Suloara'.
2. Exchange rate in June 2017: 1 GBP = 17,250 IDR; 1 USD = 13,380 IDR.
3. In the past, the creation of a ceremonial coffin (*rapasan*), made from a hollowed-out tree trunk, also involved a series of rites with animal sacrifices, marking the different phases of their manufacture. Kruyt (1924, 141) indicates that, before felling the tree (selected to make the coffin), a chicken is sacrificed to the 'spirit of the tree'; then, a dog is sacrificed once the trunk is hollowed out; eventually, a buffalo is sacrificed to consecrate the completed coffin. More generally, the presence of transitional 'steps' or 'joints' (with prescriptions of animal sacrifices) to mark and articulate the different phases of a single ritual process is typical of Toraja ceremonies: such a conception prevails for the ritual process of creating a tomb, a coffin or a house, but also (in a more elaborate form) for the performance of the funeral ceremonies (Koubi 1982, 37–39; Waterson 2009, 374).
4. *Massabu* rites are also performed for the consecration of newly made ceremonial items in Tana Toraja, including the *tau-tau* effigies of the dead, or the *saringan* palanquins used at funerary processions (Koubi 1982, 161, 165; Nooy-Palm 1986, 40, 42, 85, 196, 200, 246).

Chapter 5

Using a *liang pa'*: burial and post-burial rituals

In Chapter 1, we have presented the main aspects of the very elaborate Toraja funerary ceremonies, in particular their complex temporality (from the person's biological death to their phased funerals) and spatiality (from the *tongkonan* house to the *rante* and cemetery). In the present chapter, we take a closer look at the ritual activities directly related to the *liang pa'* cemeteries, with a focus on their material implications: *i.e.*, how these activities result in the creation of a long-term material record at cemetery sites. These activities involve various material elements that are part of the funeral ceremony and eventually deposited inside (*e.g.*, bodies of the deceased and associated material culture) and outside (*e.g.*, animal bones, material offerings) the rock-cut tombs. Following the logical sequence of the ritual, the chapter begins with a discussion of the preparation of the body of the deceased prior to burial and continues with a look at the ceremonial procession of people, animals and objects to the tomb. It then discusses the placement of the corpse inside the monument and the deposition of artefacts and substances following the closure of the tomb. We also address the various ritual activities that occur months and years after burial, such as visits and offerings at the tombs, and the *ma'nene'* ritual. The chapter concludes with cases of secondary burials (relocation of burials).

Body preparations prior to burial

In Tana Toraja, a corpse receives various types of treatments immediately after the biological death of the individual. However, the number of treatments and their degree of elaboration depend on the rank of the funeral, and therefore on the deceased's social class. In the past, bodies of slaves and commoners were simply washed, wrapped in cloths, and buried into the ground within the same day. By contrast, bodies of nobles and wealthy commoners were kept inside their houses and subjected to complex treatment processes of mummification over months and years until the funeral ceremony could be held, after which they were eventually deposited into distinctive types of tombs such as *erong* sarcophagi, rock-cut tombs or house-tombs (Kruyt 1924, 168). The complexity of noble body treatments, and their associated rituals, were studied in the 1910s by Albert Christian Kruyt (1924), and, in more details, in the 1970s, by Jeannine Koubi (1982) and Hetty Nooy-Palm (1986). These ethnographic works will be our main sources of information for this chapter. Although some aspects of these practices have changed since the time of their documentation, for instance the incorporation of chemical preservatives in the treatment of corpses, the old tradition of wrapping bodies in layers of cloths is still in use today (Waterson 2009, 379–387).

Before we discuss the details of traditional body preparations, let us reflect on their purpose and social signification. The immediate, prosaic purpose of mummifying the corpses is to buy time. Toraja noble funerals represent a high expenditure and a complex organisation, which requires several months or years for the family to gather the finances and to plan the practicalities. From the perspective of the traditional religion (*aluk to dolo*), the treatment of the body, together with the multiple rites that compose the funerals, has direct implications on the successful transition of the deceased from the world of the living to the world of the dead. Since the seminal work of Robert Hertz (1907), several studies in South-East Asia have shown how death rituals, as a social process of transition, involve processing the bodies of the dead. Defleshing, for instance, is commonly used to speed up the passing of the dead to the realm of the ancestors (Bloch and Parry 1982). In Tana Toraja, however, the intention of the body preparation is not to alter but to preserve the physical integrity of the body. As highlighted by Roxanna Waterson (2009, 206–207), maintaining the physical integrity of the body and, in particular, the skeletal assemblage, is a necessary condition for a dead person to become an ancestor (see *ma'nene'* ritual below). Individual bodies are tightly wrapped within multiple layers of cloths not only to preserve the bones, but also to keep the bones together and in their anatomical position (articulated skeletons) and to prevent bones of different individuals from getting mixed up within the collective rock-cut tombs. The logic of maintaining an individual's bones together also explains why Toraja people abhor the idea of cremation, which annihilates the physical integrity of individuals (Waterson 2009, 380).

For this reason, it is perhaps important to stress that Toraja mortuary practices should be regarded as primary burials rather than secondary burials. Although bodies go through a series of various treatments, and over a prolonged period over months or years, the bodies are materially preserved and kept inside the house; they are not buried anywhere and do not go through processes of fragmentation (defleshing or cremation) before they are eventually deposited inside the rock graves. Conceptually, moreover, individuals are not considered dead but only 'ill' or 'sleeping' during the mummification process, until the funeral begins.

Now, let us describe the body treatments. After the biological death of the individual, the body is first washed and massaged with coconut oil. The massage is also aimed at expulsing excrements from the body. This is done by the close relatives of the dead. The corpse is then dressed in fine cloths, adorned with ornaments and regalia (*e.g.*, kriss dagger), and is placed in a seated position within the middle room (*sali*) of the *tongkonan* house, where relatives and acquaintances can come visit and bring offering to the deceased (at this stage still regarded as 'ill'). In the case observed by Kruyt (1924, 139–140), the openings of the body (mouth, ears, nose, anus) are sealed with pieces of kapok, a species of wood (*Ceiba pentandra*) that is traditionally associated with death in Tana Toraja.

After a few days, once all relatives have had a chance to come pay respect, the body is brought back to the south (back) room (*sumbung*) of the *tongkonan*. At this point, the body no longer presents post-mortem rigidity. It is laid down on the floor and stretched out. Ornaments are also removed from the body, but it remains dressed in clothes. Sometimes, it is left adorned in jewellery. The body is tightly wrapped in bandages made of cotton, kapok fabric and/or pineapple fibre (*pondan*), which are aimed at absorbing the first putrefaction liquids. The wrapping gradually becomes dry and hard. This preliminary wrapping operation is made by the *to mebalun*, a slave (or person with slave origins) in charge of preparing dead bodies for funeral ceremonies.

The wrapped corpse is then placed inside a temporary coffin called *rapasan* (from *rapa'* 'to rest, to be silent, to stop speaking'): this is where the 'ill' individual will 'rest' until the funeral begins. The coffin is orientated east–west, reflecting the body position of living people when they sleep inside the house. The coffin is made from the hollowed-out trunk of a kapok or a *kamiri* tree (*Aleurites triloba* or *moluccanus*), two species of wood associated with death and therefore not used in the construction of houses or domestic furniture. Once completed, the wooden coffin resembles a rice mortar (*issong*). A *rapasan* is used for multiple funerals. When not used, it is hung against the western longitudinal beam of the *tongkonan* (van der Veen 1940, 505) or is kept in a cave in the vicinity; it is brought from there to the village and it will be returned there after the end of the funeral ceremonies. The use of a temporary coffin appears to be a very old tradition in Tana Toraja: Roxanna Waterson (2009, 11) points out that its use was reported by a Portugese traveller as early as in 1544.

Once the body is placed inside the temporary coffin, a bamboo pipe is inserted through a hole at the bottom of the coffin, and through the layers of cloths and bandages around the body to drain down the putrefaction fluids. These fluids are collected into a small pot, which will later be buried together with the deceased inside the rock-cut tomb. This process allows the body to dry out and to become mummified. According to Roxanna Waterson (2009, 380), the wrapped body is sometimes 'smoked' (*dirambu*) with a mixture of aromatic plants, although this seems to be intended to help cover the smell of the decomposition rather than to actively contribute to mummifying the body. Waterson specifies that nowadays, this complex process of mummification tends to be replaced by the injection of embalming preservatives into the body (formaldehyde). As we were often told ourselves, these preserve the body well in the short term, but result in faster deterioration of the corpse in the longer term, and therefore people that are concerned with traditional beliefs refrain from using them (see note above regarding the importance of preserving the bones in order to allow the dead to become ancestors and send blessing to their descents). The wrapped body stays in the temporary coffin for months or years until the family gathers the resources for the funeral. During this time, the *to mebalun* is in charge of taking care of the body.

At the start of the funeral ceremony, the body is moved from the south *sumbung* to the central *sali* room of the *tongkonan*. In the rite called *dibalik to mate* ('the dead has been turned': Koubi 1982, 65–66), the body is rotated to a north–south orientation, which formally signifies the death of the individual (hence no longer considered 'ill'). At the funeral of princely nobles (*puang*), the bandages previously wrapped around the body of the deceased are entirely removed. The body is then brought into a squatting position, bound in this position, and rewrapped in a new series of layers until it becomes again a sturdy cylinder of cloth (Koubi 1982, 65–66; Nooy-Palm 1986, 282–283). For other noble funerals, however, it seems that the original wrapping is left untouched and that the body stays in its original stretched position. At this stage, multiple layers are wrapped around the body, this time with cloths (not bandages), until the body gets hidden within a large cylindrical cocoon. This is the first *mebalun* rite, which concludes with covering the cylinder-like body with a single, elongated piece of fabric and stitching it to itself along its length; then, a circular piece of cloth is sewn in place at both ends. The body is then placed back again into the *rapasan* temporary coffin during the interval that separates the first and the second stages of the funeral ceremony.

During the second phase of the funeral ceremony, the wrapped body is taken out of the temporary coffin and an additional layer of cloth is sewn around it. Ovoid wooden plaques are placed at both ends of the cylinder (wrapped body), and a final piece of red cloth is added all around it (contrasting with the white cloths used for wrapping the body up to this point). The cylinder-like body then has a ridged top (see Fig. 1.11; Pl. 8). This last addition of cloths is the second *mebalun* rite. In the next stage of the ceremony, the wrapped body is decorated with beads and motifs made of gold leaves that are applied onto the outer red cloth. A sun-like (*barre allo*) motif is placed on each of the ovoid ends of the cylinder, while other motifs are applied on the longer sides. At very high-ranked funerals, coins and krisses as attached to the red layer. The body is eventually taken outside the house, where it will join ceremonial artefacts (discussed below).

Cloth wrapping is the main traditional body treatment in Tana Toraja. Bandages are used for the first layers around the body, then cloths are used to create a cylindrical cocoon. The material used in this process consists partly of cloths belonging to the deceased, in addition to ones donated by the family and relatives shortly after the death of the individual (Crystal 1985, 138). The *to mebalun* is in charge of the lengthy process of assembling, wrapping and sewing, but it is the responsibility of female relatives (called *to mangria sampin* during the funeral) to decide which cloths should be used at the various stages of the process. According to Koubi (1982, 80), body ornaments are sometimes inserted between the layers of cloth during the process. Cloth fabric itself was likely an expensive material until recently, and its use in burial practices can be considered as an index of wealth. The number of layers and the thickness of the cloths reflect the rank of the deceased individual. As is the case for house and tomb sizes and decorations, there may have been an increase over time with regards to the average thickness of the wrapped body cylinder. For instance, at some cemeteries we visited, there were old tombs with doors that were missing or cracked open, which enabled us to see relatively ancient examples of wrapped bodies (see Fig. 5.7, below). The layers of cloth around these bodies seem to have been much thinner than the more recent examples we saw (Fig. 1.11; Pl. 8). Alternatively, the bodies wrapped with thinner cloths in older tombs may represent burials of lower ranks.

During the process of applying cloths, it is important for all the cloth layers to be very tightly wrapped and sewn together. According to Kruyt (1924, 143), this was referred to as 'making a tree trunk' (ma'batang[1]). According to Nooy-Palm (1986, 283), the aim is to 'attain a smooth cylinder, so sturdily constructed that, should the need arise, it can stand by itself'. According to Kruyt (1924, 145), bamboo slats were also used to form the cylinder around the body. The rigidity of the shrouded body is important from a practical perspective, since in the latest stage of the funeral, it will be carried on a bamboo ladder up to the cliff and inserted into the rock grave by a small group of men. Some people nowadays prefer to use wooden coffins instead of the traditional *mebalun* wrapping, perhaps as a result of Christian influence. Some of these coffins are made by skilled artisans: they imitate the ovoid-cylinder shape of the traditional wrapped bodies and are lavishly decorated with wood-carved traditional motifs – somewhat resurrecting the lost tradition of the wooden *erong*.

Ceremonial artefacts: from production to procession and deposition at the tomb

Funeral ceremonies in Tana Toraja involve many kinds of artefacts, from the family heirlooms displayed around the corpse inside the house, to the drums used during sacrifices in the courtyard outside (Koubi 1982; Nooy-Palm 1986). However, only a few of them are created specifically for the funeral and follow the dead to its final resting place on the rock cliff. This section focuses on four types of objects that are directly related to the body of the deceased and which are eventually deposited in front of the rock burial: the *saringan* palanquin, the *tau-tau* effigy, *dulang* wooden plates and personal accessories.

Saringan *palanquins*

The *saringan*[2] is the wooden structure that is used to carry the wrapped corpse from the hamlet (*tondok*) to the ceremonial plaza (*rante*) and then to the rock-cut tomb cemetery as part of the funeral ceremony. A *saringan* is built specifically for one funeral and cannot be reused for any other funeral. Its construction is part of the funeral ceremony and takes place at the beginning of the second phase of the funeral. On the last day of the funeral, at the end of the procession, the palanquin is abandoned in front of the tomb at the bottom of the rock face.

A *saringan* has two main parts, a litter and a roof (see Fig. 1.11). Each is built separately and they are assembled together shortly before the departure of the procession, once the corpse has been placed on the litter (Koubi 1982, 161; Nooy-Palm 1986, 247). The litter is a wooden, rectangular structure that is the basis of the palanquin. Its sides are made with decorated planks, and its top is the platform, the dimensions of which correspond to the size of the wrapped corpse (*e.g.*, 0.50 × 1.60 m). Six or eight vertical posts are fixed on the edges of the platform (the number of posts reflecting the rank of nobility, as per *alang* rice barns). The posts support both the litter and the roof structure above it. The roof is an imitation of a *tongkonan* house or an *alang* rice barn in miniature, with its decorated wooden walls and its long, curved roof covered in bamboo and corrugated metal or plastic sheets. Long and large bamboo segments are fixed under the platform so the *saringan* can be carried by people during the procession (Fig. 5.1). The bamboo is later removed once the *saringan* has arrived at its final destination at the cemetery.

Figure 5.1: Funeral procession in Lempo (Sesean Suloara'): a deceased individual is carried across rice paddies by relatives in a saringan *palanquin to its rock-cut tomb (photo: G. Robin).*

This type of palanquin is only used at funerals of noble rank, more specifically at funerals at which at least ten water buffalos are sacrificed. *Saringan* are expensive too: in 2017, it costed about 10,000,000 Rupiah to build one (*c.* 580 GBP or 750 USD[3]). Lower-ranked funerals use simpler, non-decorated and unroofed bamboo litters.

Saringan are made with special types of wood, such as the *kamiri* (candlenut, *Aleurites moluccanus*), *nato* (*Michelia celebica*) or kapok tree. They are decorated with a variety of motifs that are also found on traditional houses. The wooden parts of the *saringan* are first blackened with charcoal. They are then engraved with a knife by specialized woodcarvers. The carvings are finally painted with red, white and yellow colours by apprentice children working with the adult woodcarvers.

Creating a *saringan* is more than just a technical process. As is the case for constructing a house or cutting a tomb (Chapter 4), it also requires particular rituals, which are carried out at each stage of the construction process of the palanquin, and are part of the funeral ceremony (Koubi 1982, 92, 157, 161; Nooy-Palm 1979, 261; 1986, 195–196, 246). The first ritual is called *manglelleng sarigan* ('to cut [the wood for] the *saringan*') and involves the sacrifice of three dogs, a chicken and a pig (Koubi 1982, 157) or the sacrifice of a chicken, and the blood of a dog (a cut is made behind its ear, which is rubbed with cloth, after which the dog is donated to the *to mebalun*) (Nooy-Palm 1986, 196). The sacrifices are executed by the carpenter and one of his assistants. At the highest-ranked funeral (*dirapa'i*), the blackening of the wooden parts of the palanquin involves a ritual: *mangrambu bulisak* ('to blacken the wood chips with smoke' – a pig is slaughtered). This optional, high-ranked rite takes place after the wood parts are cut, and before they are assembled.

The second ritual corresponds to the assembling of the wooden parts to form the *saringan*. It is called *ma'rapa' sarigan* ('to make touch' or 'to assemble the *saringan*': Koubi 1982, 161) or *mesarigan* (Nooy-Palm 1986, 195) and consists of sacrificing a pig. Once the *saringan* is assembled, a third rite takes place for its consecration. This is called *massabu sarigan* ('to consecrate the *saringan*') or *mangrara sarigan* ('to oint the *saringan* with blood'). The consecration rite requires the sacrifice of a pig. It is carried out before the wrapped body is lifted onto the palanquin.

Once the *saringan* is completed and consecrated, it is ready to be used for the procession. The wrapped body is placed on the litter. According to Kruyt (1924, 144), the number of straps used to fasten the corpse to the litter reflects its rank, with three straps for a *to makaka* noble and five for a *puang*. The *saringan* is transported with two or more long bamboo poles that are inserted under the litter and carried by men. According to Nooy-Palm (1979, 261), 'some twelve to twenty men are needed to carry the *sarigan* to the *rante* During the funeral we observed in Lempo in 2017, 12–14 men carried the *saringan* (Fig. 5.1). The wrapped corpse is carried in the *saringan* from the *tondok* to the *rante*, and from there to the cemetery. Once it has arrived at the cemetery, the body is taken out of the *saringan* and lifted up to the grave. After the entombment, the *saringan* is left near the entrance of the tomb, either at the bottom of the cliff or at the top of the boulder in which the tomb is cut. Henceforth, the *saringan* should no longer be moved or even touched. It will stay there and decay progressively. During our visit in 2017, we observed *saringan* in various states of deterioration.

Tau-tau *effigies of the dead*

Tau-tau (literally 'little person' or 'like a person') are wooden statues representing dead individuals. A *tau-tau* is created at the occasion of a high-ranked funeral ceremony, and specifically represents the person celebrated during the funeral. The effigy plays a part in the ceremony and procession and is eventually deposited in front of the tomb. This type of ceremonial artefact is mainly associated with *liang pa'* tombs but is also found associated with other tomb traditions in both Sa'dan and Mamasa regions. *Tau-tau* are found at *erong* cemeteries (*e.g.*, Tampangallo and Marante), which indicates that their use pre-dates the *liang pa'* tradition. They are also found in house-like tombs (*patane*).

This intriguing ritual apparatus is a central element in Toraja funeral ceremonies and has received particular attention from anthropologists as part of dedicated studies (Volkman 1979a; Crystal 1985) or broader descriptions of funeral ceremonies (Nooy-Palm 1979, 261–263; 1986, 213–214, 244–246; Koubi 1982, 157–165). Here, we will only summarise the key elements of its composition, manufacture and uses, and emphasise its connection with the dead bodies of the individuals and with the cemetery site.

Tau-tau are made exclusively from the wood of the *nangka* tree (jackfruit, *Artocarpus heterophyllus*), a particularly highly valued type of hardwood used in Tana Toraja for specific elements of house constructions and which is associated with durability. When aged and lubricated with coconut oil, the wood takes a darker colour which is deemed to very closely resemble the skin of Toraja people. Each effigy is gendered. It is made specifically for a deceased person and should bear a close representation to the person (Figs 5.2–5.4). This is achieved by the carving of the face (a more or less detailed portrait of the person) and by cloths added onto the effigy that belonged to the deceased individual. The effigy is typically from two-thirds life-size to life-size compared to its human model. It is normally made of several wooden parts, chiefly the body trunk, the head, the arms and the legs, which can rotate and thus enable the effigy to be animated during the ceremony. However, several of the older *tau-tau* we observed at *erong* cemeteries seemed to be monoxyle. They are typically standing or are otherwise sometimes in a seated position. Their designs vary but important elements that are constant include their wide-open eyes (often made from a white material), and their hands (often only the right one) that are lifted and open facing upward, as a way to express that the dead are constantly addressing the living and asking for attention and offerings. The *tau-tau* indeed are not just representations of deceased individuals, they are believed to be inhabited by their soul or spirit (*bombo*[4]). The effigy is referred to as *bombo dikita* ('the soul of the dead which can be seen'). The main function of the *tau-tau*, therefore, is to presence the deceased person both during the ceremony and at the cemetery site, thus enabling different types of interaction with her/him.

Tau-tau are only made for the highest-ranked (*dirapa'i*-type) funeral ceremonies, which entail the sacrifice of at least 36 buffaloes. They are therefore associated with high nobility and prestige. Another type of effigy, called *tau-tau lampa* or *bate lepong* can be used during lower-ranked funerals. This is a temporary effigy made of bamboo wrapped in cloth, which is only used during the funeral and is dismantled when the ceremony is finished. The wooden *tau-tau*, on the contrary, are made from durable material and are aimed at being displayed at the tomb site for generations after the funeral takes place. They are commissioned by the family sponsoring the funeral and are manufactured by specialised wood carvers (called *pande tau-tau, pande kayu* or *to ma'tau-tau*) during the duration of the ceremony. As per *saringan* palanquins, specific rituals mark the different steps in the process of creating a *tau-tau*. The *manglelleng tau-tau* rite ('cutting [the wood for] the *tau-tau*'), which involves the sacrifice of a chicken, is performed when a living *nangka* tree is felled. The next rite, called *ma'tau-tau* ('to make the *tau-tau*'), takes place during the carving of the body parts from the hardwood and involves the sacrifice of a pig and a chicken. This rite is performed by the wood carver to solicit help from the spirits of nature and the ancestors in

Figure 5.2: Tau-tau *effigy of Ne' Lai' Ambun, placed under a rice barn as part of her high-ranked funeral ceremony in 1937 in Tadongkon (Kesu) (photo: unknown author, public domain, Leiden University Libraries).*

fashioning the effigy. Once the body parts of the *tau-tau* are complete, the rite called *ma'rapa' tau-tau* ('to assemble the effigy') can take place, with a pig sacrifice carried out by the traditional priest (*to minaa*) who supervises the entire funeral ceremony. The effigy is then assembled, dressed up, and displayed on the rice barn platform (Fig. 5.2) located in front of the *tongkonan*, inside which lays the wrapped body of the deceased. Nooy-Palm (1986, 196) mentions an additional rite at this stage, *manglassak tau-tau*, which takes place when the effigy is given genitals, and requires a pig sacrifice. The *to minaa* also performs the final rite for the *tau-tau*, *massabu tau-tau*, which marks the consecration of the effigy and involves the sacrifice of one more pig. The effigy is now ready for its ritual use.

After the *tau-tau* is completed and consecrated, it is brought inside the *tongkonan* to the room where the wrapped body is kept. The effigy is placed in sleeping position

5. Using a liang pa': burial and post-burial rituals

Figure 5.3: Rock-cut tombs and tau-tau *effigies in rock-cut galleries photographed by Albert Grubauer in 1911 in Tondong Dandelolo near Makale (Collection Wereldmuseum Coll.nr. RV-A83-1-191).*

Figure 5.4: Tau-tau *effigies in a rock-cut gallery at the cemetery of Suaya (Sangalla) (photo: G. Robin).*

beside the deceased and is still considered inanimate at this stage. Later, when the body of the deceased is turned from an east–west orientation to a north–south one inside the *sali* room of the house (marking officially the death of the individual), the *tau-tau* is 'awakened' (*ditundan*) and placed upright beside the body, facing also to the south (in the direction of Puya). According to Crystal (1985, 134), this rite means that the soul of the deceased, which had been wandering in and around the house in the previous days, has now entered the effigy, or, more precisely, is now animating the effigy. The *tau-tau* is then given food offerings for several days, before joining the procession that leaves the hamlet to the *rante* and cemetery.

The *tau-tau* leads the procession, carried in a sedan chair in front of the *saringan* containing the wrapped body. Two processions are sometimes organised, a first with the *tau-tau*, and a second with the *saringan*. During the ceremonies taking place in the *rante*, the *tau-tau* is placed under the *lakkean* platform (which exposes the wrapped body) in a position from which to view the animal sacrifices. Before the procession leaves the *rante* to go to the cemetery, the *to minaa* gives a food offering to the effigy as part of a rite called *ma'pakande tau-tau* ('to feed the effigy').

Arriving at the cemetery and following the placement of the wrapped body inside the tomb, the effigy is placed high up in the rock face: either in the small recess at the entrance of the *liang pa'*, or in dedicated rock-cut galleries with a balustrade, as in Lemo and Suaya (Figs 1.1; 1.7; 5.3; 5.4). The display of the *tau-tau* on the rock cliff is twofold: as a social statement, they ensure that the names and very high status of certain deceased individuals will be remembered (thus conferring prestige on their family); they also remind the living that the ancestors are watching them and that traditional proscriptions should be respected. In recent decades, unfortunately, many *tau-tau* displayed in isolated cemeteries have been robbed to be sold illegally in native art markets. As a consequence, most Toraja families have removed the *tau-tau* from the cemetery cliffs or have placed them inside the rock-cut tombs, as in Batu Lappa' where the rock-cut *tau-tau* gallery is now empty (Pl. 15).

Dulang *wooden bowls and other personal accessories*
The *dulang* is a wooden bowl or dish on a pedestal. In the past, this special type of vessel was only used by the nobility (Nooy-Palm 1979, 53; Waterson 2009, 410), and was made from long-lasting hardwood, such as *nangka* (jackfruit). It had a collective use, for instance, to serve rice in a shared meal within the house, or at ceremonial occasions, such as house or fertility rituals (Nooy-Palm 1986, 97, 157–158). It is also traditionally used at ceremonies to receive offerings intended specifically for the ancestors (Waterson 2009, 255, 340). However, some *dulang* were also used as personal dining plates by individual nobles; therefore, specific *dulang* were associated with specific individuals. As such, they took on a role in funeral ceremonies: a deceased's own *dulang* was placed on the ground next to one end of the wrapped body, together with food offerings in other containers (Nooy-Palm 1986, 188). Later on, during the procession from the village to the ceremonial plaza (*rante*), the *dulang* is held by the *to mebalun* (together with the personal hat in the case of a deceased woman) (Koubi 1982, 178, 194–195). The wooden bowl is either deposited on the stone bench of the recessed entrance or on the ground at the bottom of the rock face. The practice of depositing *dulang* in front of noble tombs is very ancient, as these can be seen today at old *liang pa'* cemeteries such as Pana' and Parinding (boulders 9 and 13), as well as at *erong* cemeteries such as Londa (Fig. 5.5). It is not clear whether, once placed at the tomb entrance (or at the foot of the cliffs), *dulang* were regularly 'used' to receive food offering over time (in the same way they are used at house-rites to receive food for the ancestors), or whether they were left untouched. Volkman (1985, 145) indicates that at *ma'nene'* rituals (see below), relatives come and place offerings of betel in them. Today, *dulang* seem to have been replaced by offerings of water bottles placed in front of the tomb door (see below).

Other personal belongings of the dead can be placed at the entrance of the tomb as part of the burial ceremony. In the case of female burials, the finely woven bamboo sun hat (called *sarong*: Waterson 2009, 340) is also carried by the *to mebalun* during the procession to the cemetery, and is eventually hung on the door of the rock grave (see Fig. 3.8). This is normally the hat of the deceased woman, although Kruyt (1924, 149) mentions that this type of hat was worn by the widow of a deceased male individual during the ceremony and was later deposited in front of her husband's tomb. Nobele (1926, 51) indicates that both men and women wear such hats (called *sarong bunga* for a man, and *sarong manic* for a woman), which were carried out by the *to mebalun*

Figure 5.5: Left: a dulang *wooden bowl in a private collection in Tonga Riu (Suloara'); right:* dulang *bowls deposited under the entrances of ancient* liang pa' *rock-cut tombs in Parinding boulder 9 (Sesean) (photos: G. Robin).*

during the procession and hung on the tomb door after the burial. During our field visit in 2017, we also noted other personal items such as umbrellas at both old and recent *liang pa'*.

Some high noble burials (either male or female) are also associated with a long piece of fabric that is attached to the wrapped body and extended from the body over the closed door and outside the tomb, hanging down from the tomb on the cliff as an external sign of status (*e.g.*, at Suaya; see Fig. 1.7). This cloth can be either red (*lamba lamba*) or white (*duba duba*). Prior to burial, it is attached to the *saringan* and carried by participants during the procession (Koubi 1982, 178; Nooy-Palm 1986, 303).

Opening of the tomb and entombment

Liang pa' are family tombs and only members of the family can be buried within or, more specifically, only descendants of the founder of the tomb can claim a burial right in the monument. Individuals can have connections with many different kinship groupings and therefore could theoretically claim burial rights in several *liang pa'*. However, in practice, the choice of the final resting place is negotiated collectively. Some *liang pa'* (i.e., *tongkonan* affiliations) are considered more prestigious than others; therefore, it can be judged inappropriate for a certain person (genealogically too distant, or in too low of a social position) to be buried in them. Conversely, it is inappropriate to bury an individual from a high nobility background in a *liang pa'* belonging to commoners or lower-ranked nobles. Bypassing this collective negotiation and placing a body in a 'wrong' *liang pa'* is perceived as a significant offence. This is discussed in detail in an excellent article by Roxana Waterson (1995), and we will not address the complexities of this matter here (examples of kinship connections to specific *liang pa'* within large cemeteries will be presented in Chapters 6 and 7). Instead, we will focus on the general ritual activities and their material consequences as they occur during the final stage of the funeral process.

The placement of the body inside the tomb has many noteworthy implications. Is there any ritual associated with this key moment? Who is in charge of carrying the dead body up the rock face and inserting it into the rock tomb? How many bodies can a rock tomb have and what happens if a tomb is full? Does the position of the bodies inside the tomb matter?

A deceased individual can be buried in either a newly built *liang pa'* that has never been used before, or in a tomb that has already been used. The former situation is probably more common today than in the past, considering the increasing number of new tombs that have been cut in recent decades. In this particular case, the tomb is normally left empty and open after the completion of the construction work and is consecrated only shortly before its first funerary use (see Chapter 4). The second burial scenario involves a *liang pa'* that was used previously and therefore contains previous body depositions and is closed by a wooden door. Interring the deceased in a previously used tomb was likely the most common situation in the past, and remains common today. In these cases, the opening of the tomb for a new burial deposition is perceived as a disturbance for the dead already interred in the tomb and therefore requires specific rituals. The rituals held for these types of burials are called *ma'peliang* ('to place into the *liang*') or *meaa* ('to bury') (Koubi 1982, 209; Nooy-Palm 1986, 367; 1988, 99–100). The opening of the tomb itself requires the sacrifice of a pig. This can be done on the day of the entombment, or several days before, in order to clear the vegetation around the tomb and clean and prepare the chamber, without removing or displacing the previous burials (Waterson 1995, 209). Ancient *liang pa'* are often located in high positions in cliffs. Accessing the *liang pa'* involves the building of a long bamboo ladder or a scaffolding, which requires some time ahead of the entombment.

Traditionally, when a new body is placed in a *liang pa'*, every person who has an ancestor buried in the tomb has an obligation to attend the entombment ceremony and to offer betel to the dead (Waterson 1995, 212). Attendance at this ceremony is not restricted to members of the *tongkonan* group. According to our guide Samuel Palangda', the entombment ceremony is open to anyone in the community, and even foreigners. Only pregnant women were prohibited from attending this ceremony in the past, as it was believed that pregnant women would attract bad spirits to the *liang pa'* site. As Samuel Palangda' stated, that was because a pregnant woman 'smells good'.

One important reason why burial ceremonies were open to all, even children, was to ensure that people would develop knowledge concerning which family and *tongkonan* is associated with what tomb. If one sees many burials in a location over a long period of time, they should come to know what burial location(s) are reserved for their own *tongkonan*. This would help ensure that corpses are not accidentally interred in the wrong *liang pa'*. Such corpses placed in the wrong *liang pa'* are called *to pusa*. Although there is no specific penalty for placing a corpse in the wrong *liang pa'*, the family of the deceased is obligated to move the corpse to the correct *liang pa'* location when this occurs (see also Waterson 1995 for a discussion of burial rights and the offence constituted by the placement of a deceased in the wrong tomb). According to Samuel Palangda', when a *liang pa'* tomb is opened for a burial deposition or a *ma'nene'* ritual (see below), leaves from the induk palm tree (*Arenga pinnata* – same tree used for making palm wine) were traditionally placed around the edges of rice fields near the *liang pa'* site. This was done to protect the crops from bad spirits that may be attracted to the area.

Grubauer (1913, 258) and Nobele (1926, 51–52) provide useful descriptions of the *meaa* ritual as it was carried out traditionally in the past. The wrapped corpse is taken out of the palanquin and carried by male members of the deceased person's family to the tomb entrance. As we could observe for ourselves at a burial in Lempo, this process is done relatively quickly and without particular ceremony (Pl. 8; Fig. 5.6). Once the corpse has reached the entrance of the tomb, it has traditionally been the responsibility of the *to mebalun* to insert the body into the burial chamber, together with the putrefaction fluids (contained in a ceramic vessel or bamboo tube). The body is placed head first, directly on top of the previously interred body depositions. The *to mebalun* then closes the tomb's wooden door after pulling a segment of the *lamba lamba* cloth (an extremity of which is attached to the wrapped body) out over the top of the door shutter. Then, he hangs the bamboo hat of the deceased on the front of the door and places the *tau-tau* effigy and *dulang* bowl on the stone bench in front of the door. Once the entombment is complete, animals are slaughtered (a buffalo and several pigs in the case of a *puang* burial) in front of the tomb, as an offering to the soul of the dead and the *deata* (spirits of nature). Kruyt (1924, 143) specifies that the meat is subsequently cooked there immediately following the slaughter, and partly consumed by the participants, partly left at the tomb site (no part of this meat can be taken away).

How many bodies can be placed in a rock-cut tomb? The main factor here is the volume of the chamber, which varies from site to site, and across the two main generations of *liang pa'* (see Chapter 3). Brisbois and Douvier (1980, 116) report that up to 12 wrapped bodies can be held inside chambers measuring 1 × 1 × 2 m (*i.e.*, older *liang pa'* generation). During our 2017 visit, we observed some old, no longer maintained *liang pa'* with deteriorated doors through which it was possible to see depositions of wrapped bodies (*mebalun*) inside the tombs (Fig. 5.7). In these instances, tombs contained two bodies laid side by side, and up to four bodies piled up in two levels. At the cemetery of Salu Liang, another disused tomb (with a deteriorated wooden door) had no *mebalun* visible but only commingled human remains,[5] from which ten skulls could be counted, giving an indication of the burial capacity of this type of tomb.

Figure 5.6: Placement of a wrapped body inside a liang pa' rock-cut tomb in Lempo boulder 18 (Sesean Suloara') (photo: G. Robin).

112 Ethnoarchaeology of Rock-cut Tombs

Figure 5.7: Burials inside old liang pa'*, whose door shutters are missing. Parinding boulder 4 (Sesean), top, and Salu Liang (Malimbong Balepe'), bottom left, have three wrapped bodies (*mebalun*) resting on top of commingled human bones. Another tomb at Salu Liang, bottom right, has a very deteriorated door and contains only comingled bones, suggesting it is one of the oldest tombs in the cemetery (photos: G. Robin).*

Information on people buried in the cemetery of Sele (see Chapter 7), also indicates that up to ten people were buried in the older style of tombs. We can conclude that a typical old-style *liang pa'* could host up to 10–12 individuals, although we have been told that some of the old tombs at Lemo were designed to receive only one body.

What happens when a *liang pa'* is full? It is forbidden to remove bones from a *liang pa'* in order to make room for a new burial, as was confirmed to us by *to minaa* Paulus Kondosara. There are therefore two main solutions to create additional burial space. The first is to create a new tomb. For instance, our guide Samuel Palangda' is affiliated (from his father) to *tongkonan* Mabarre' which has two *liang pa'* tombs at the cemetery of Batu Lappa'. The first tomb was likely built when *tongkonan* Mabarre' was created three generations ago by Palangda's great grandparents, who are buried in this *liang pa'*. The second tomb was built two generations ago and contains Palangda's grandparents who were buried there in 1982, as well as his grand uncle who was buried there in 1992. Other people had been buried in these tombs before these dates and both are now considered 'full'. As a consequence, a new tomb (in this case a concrete *patane*) was built by Palangda's brother about 100 m away from *tongkonan* Mabarre', and contained three persons in 2017 (Palangda's brother, aunt and father's first cousin).

The second option, which we found more surprising, is to wait for the natural decomposition of the bodies inside the tombs to create room for new burials. This was explicitly told to us by Palangda's father (Abraham Sulu'), for whom the second *liang pa'* in Batu Lappa' could host new burials in the future when older burials will have 'turned into dust'. The idea that room is constantly created naturally inside tombs, making them in practice never permanently full, was confirmed to us by *to minaa* Paulus Kondosara who used the expression *memangan* ('it is astonishing, wonder-inspiring') to express the fact that somehow, without a clear explanation, room is always made naturally inside tombs over time (see also Kruyt 1924, 162). In fact, in the old open tombs mentioned above, it was possible to see a layer of comingled bones mixed with debris of fabric (*i.e.*, deteriorated *mebalun*) at the bottom of the chamber's 'stratigraphy', under the more recent *mebalun* depositions (Fig. 5.7). This suggests that older depositions gradually break down and deteriorate, creating space for new burial depositions over time.

More recent *liang pa'* are more spacious than older ones, with a chamber measuring 2 × 2 × 2 m. We never came across examples of such tombs being full, likely due to the fact that they have large capacity. Moreover, more *liang pa'* tombs are built today than in the past, due to higher economic capacities of people (within and beyond traditional nobility). This has resulted in a higher number of *liang pa'* tombs available for communities and families. Therefore, each tomb tends to be used by a smaller group of people (*e.g.*, a nuclear family), while in the past individual *liang pa'* would be used by a whole *tongkonan* group or multiple generations. Typically, people today prefer to create a new tomb if they have the resources, rather than opting for the older ones packed with previous body depositions. Indeed, several tomb commissioners we spoke to in 2017 stated they preferred not to use the existing *liang pa'* of their *tongkonan* kinship group, and wanted a new, separate tomb for themselves and their closest relatives.

Dead bodies deposited in both past and present *liang pa'* are tightly packed and piled up against each other inside the rock chamber. Such a physical proximity between the dead is a matter of concern. Waterson emphasises how Toraja aristocratic families do not want to see individuals of different ranks being mixed up inside tombs. A more recent concern is from Christian Toraja, who do not want to be buried together with their pagan ancestors and therefore prefer to be buried in newer tombs, such as *patane* (Waterson 1995, 207). Are there any strategies to prevent the mixing of bodies inside the tombs, other than placing them into separate tombs? Koubi (1982, 196) states that *liang pa'* are sometimes compartmentalised, so that the most important members of the family can be separated from the others, although she does not specify how. According to our survey, the only way to do this is to create an upper recess off the chamber space, which is normally reserved for the sponsor of the tomb (Chapter 4). According to Pak Saipan, one of the stone workers we spoke to in 2017, it is possible to divide a tomb into two spaces, using an internal wall made from wood of the induk palm. This can be done to separate people from two different families (such as brothers and their respective families) that are linked to the same *tongkonan*, for example. In Bori', we viewed an example of this type of arrangement in the case of a tomb that included both the husband and the wife separated within the same tomb (Fig. 5.8).

There are several references in the literature to cases where slaves were buried with their noble masters inside *liang pa'* tombs. According to Kruyt (1924. 162–163) and Nooy-Palm (1979, 259–260), some tombs had two consecutive chambers, one at the front (for slaves) and one at the back (for the family members). Not all the slaves owned by a family were buried in the front chamber. Only important members of the family could select one slave to be buried with them inside the tomb, to ensure that they continued serving them in the afterlife. Slaves selected for this function were called *pelilli' liang* (from *lilli'*, 'underlay'). If their masters died before them, they would not be killed; they would only join their masters in the tomb after their own death. In the meantime, clothes belonging to this slave were placed under the body of the deceased master.

According to Keers (1939, 207–208), van der Veen (1940, 309) and Koubi (1982, 196), bodies of slaves used to be buried under the bodies of their masters, with the same idea of the former serving the latter in the afterlife. The slave bodies were the first to be placed in the chamber of newly hewn tombs and were 'regarded as foundations' (*dipopenlilli'*), i.e., the first layer supporting the subsequent stacked depositions of noble bodies. If the slave survived the masters, a mat or a layer of clothes belonging to the slave were placed underneath instead. During our 2017 visit, we were told of a traditional ritual called *manglilli' liang* which involved placing the clothing of a slave person on the floor of the *liang pa'*. The noble person owning the slave would be placed over the slave's clothing with the hope that the slave would serve their master

Figure 5.8: This liang pa' *in Bori' boulder 26 (Sesean) has an internal wooden wall dividing the chamber into two spaces: one for the relatives of Ne' Sea (left-hand side), the other for the relatives of Ne' Lai' (right-hand side) (photo: G. Robin).*

for eternity. Since the abolition of slavery in Tana Toraja from the late 19th century, and the gradual disappearance of the status of *kaunan* (slave) (Waterson 2009, 161), such rituals have no longer been practised. However, it still seems to be considered inappropriate to place a wrapped body or coffin directly on the rocky floor of the *liang pa'*. In the entombment we observed at Lempo, a mattress was brought on the day of the burial and placed under the wrapped body, which was the first deposition in the recently hewn tomb.

Post-burial depositions outside the tomb

Rituals and offerings take place in front of the tomb shortly after the entombment, and in subsequent months and years. The first are part of the closure rituals which mark the end of the entire funeral ceremony and the associated mourning period. The second are made at informal visits by relatives, or at special ritual occasions (*ma'nene'*).

Rituals of funeral closure involve a rite performed at the cemetery. This rite takes place a few days after the entombment and consists of animal sacrifices addressed to both the deceased and the *deata*. In the Rantepao area, it is called *umbaa kande* ('to bring food'), *ma'pakande to mate* ('to give food to the dead'), or *parundun bombo* ('to bring food to the bombo' as the 'spirit' or 'invisible double' of the dead). It consists of bringing the meat of a pig (previously sacrificed at the deceased's house) as well as vegetables to the cemetery where they are deposited at the foot of the rock face in which the tomb is located. It is performed by close relatives of the deceased (Nobele 1926, 52–53; Koubi 1982, 210–211; Nooy-Palm 1986, 185–201). In the Rembon area, the rite is called *ban manuk* ('to give chicken') and involves bringing cooked chicken and vegetables for the deceased, as well as the sacrifice of pigs (Waterson 2009, 319–320).

Once the funeral closure has been ritually completed, occasions to go again to cemeteries and view the tombs are rare, especially in the past. Traditional rock-cut tomb cemeteries were located in isolated places in the landscape, separate from daily life activities. People would not venture to cemeteries that were perceived as sacred places conducive to direct communication between the dead and the living (Koubi 1982, 208, 238), and at which 'strange occurrences were common experiences' (Waterson 2009, 365). In fact, according to traditional prescription (*aluk to dolo*), just as funerals cannot be held before harvest, cemeteries can only be visited between harvest and the planting of rice, that is between August and September each year (a period also described as the 'funeral season': Waterson 2009, 305). This proscription is due to the association between the ancestors and crop fertility. Visiting (*i.e.*, disturbing) the dead before harvest could have an impact on crop production. Today, many cemeteries in Tana Toraja have become signposted touristic sites (*objek wisata*) and can be visited any time. In certain areas where Christian influence is strong, the *liang pa'* are visited at Christmas and New Years at which time offerings, such as plastic water bottles of coffee are placed in front of the tomb. The traditional proscription, however, is still perceptible in some places. As foreigners visiting Tana Toraja in the month of June, we were not always allowed to approach remote cemeteries or to remove vegetation from the entrance of more accessible tombs while 'the rice is still growing'. This means that traditionally, Toraja people could only visit the tombs of their ancestors during a single short period each year.

Outside of special ritual occasions, such as a burial or a *ma'nene'* ritual (see below), it is possible to visit the tombs and to make offerings to the ancestors. These are normally placed in front of the tomb wooden door, on the stone floor of the recessed entrance. In the past, the main type of offering was betel nut, while today it is very common to find water bottles, cigarettes, banknotes and even crisps placed at tomb entrances (see Figs 3.9 and 3.10).

Visiting cemeteries outside the funeral season was traditionally considered as *pemali* ('prohibited'), but the worst offence was to open a *liang pa'* and steal its contents. This transgression is called *meloko* ('to steal from a *liang*'), and a person who committed it is called *to meloko*. According to Grubauer (1913, 258–259), this was the most serious offense known in the Toraja customary laws and was punished by death. Koubi (1982, 208–209) specifies that the highest of the purification rites must be performed after a robbery in a *liang pa'* to appease the ancestors. The rite is called *ma'rambu langi'* and involves the sacrifice of a pig (see also van der Veen 1940, 494–495).

Ma'nene' ritual

The *ma'nene'* ritual is a major ceremony that takes place in cemeteries and involves interactions with the bodies of deceased previously buried in the tombs (Volkman 1979b; 1985, 144–147). Its name comes from *nene'* ('grandparents' or 'ancestors'). In most cases, it consists of taking the bodies outside the tombs, changing the cloths of the mummified corpses (*i.e.*, unwrapping the bodies, giving them new clothes and rewrapping them with new fabrics), and placing them back into the tomb. If the *mebalun* wrapping is in good condition, the new cloths are simply placed inside the tomb beside the bodies. In the case of more recent burials where bodies are simply put in coffins, the latter are taken out of the tombs, opened and bodies are given new cloths. A *ma'nene'* ritual also pertains to the *tau-tau* effigies in cemeteries where they are present. The effigies are given new clothes together with, or instead of, the bodies of the deceased they represent.

The ritual varies and takes on different names across Tana Toraja. *Ma'nene'* seems to be the most common name, but others are also used locally, *e.g.*, *maro* in the Sesean region, *ma'tomatua* or *mangeka'* in Kesu', and *ma'paundi'* in Saluputti (Nooy-Palm 1986, 170–171, 203–207; Waterson 1993). This ritual is not performed everywhere in Tana Toraja. It has never been practised, even in the past, in certain areas such as the district of Rembon. In other areas, *ma'nene'* used to be done but now belongs to the past. It is still practised today in a few places in the northern part of Tana Toraja, such as in the district of Sangalla, where the *puang* cemetery of Suaya is located, and more broadly in the northern districts of Mount Sesean, in particular in Rindingallo, where it is still performed every year. At Tampangallo, an old *erong* cave cemetery in Sangalla, new burials have not been placed for generations, but the *tau-tau* are still re-oiled and given new clothes regularly today. This means that in some areas, *ma'nene'* ritual is done for long dead ancestors of particularly high rank.

The *ma'nene'* ritual is done at the end of the funeral season (August–September), before the planting of rice. Its scale and organisation vary across regions. It can be organised by a single *tongkonan* for only one person that has been recently buried, typically 3 years after the burial (a delay that ensures the *bombo* of the dead person has reached Puya, the land of the dead). Elsewhere, it takes place on a regular basis, every 5 or 10 years, and is organised on a communal basis: several *tongkonan* groups gather at the cemetery where tombs are open and several corpses rewrapped (Waterson 2009, 303–306). We were told that a *tongkonan* would normally do a *ma'nene'* ritual every year, but on a different tomb each year, on a rotating basis (so a *ma'nene'* ritual is not done for the same *liang pa'* two years in a row). *Tau-tau* effigies can also be given new clothes at the occasion of a major funeral ceremony (i.e., outside a *ma'nene'* ritual), as described by Nooy-Palm (1986, 301) in Suaya. *Ma'nene'* rituals also provide communities with an opportunity to inspect, clean and repair tombs (for instance their wooden door), while they should remain untouched during the rest of the year (Volkman 1985, 145; Waterson 2009, 207).

Traditionally, it was the role of the *to mebalun* to unwrap and rewrap bodies at *ma'nene'* ceremonies (Waterson 1993, 87), although today this is carried out by the close relatives of the deceased. The ritual involves the sacrifice of a pig in front of the tomb, as the latter needs to be opened. Offerings of betel, rice, pork, wine and cigarettes for the dead are made at this occasion. According to Pak Saipan (a stone worker we met in Bori'), a *ma'nene'* ritual also involves the planting of a *sendana* (sandalwood, *Santalum album*) tree in front of the *liang pa'* at which the ritual is held (see also Nooy-Palm 1986, 340, n. 11). After the planting, the tree must remain and be left to grow naturally; it should never be cut down as it would displease the ancestors. This tree planting echoes the *ma'bua* or *merok* ceremony (consecration of a *tongkonan*) which involves the planting of a *sendana* in front of the *tongkonan*.

The main purpose of the *ma'nene'* ritual is to renew contact with the ancestors, and to create a moment of interaction with them. At this occasion, the living demonstrate that they care for the bodies of the ancestors, by offering them food and new cloths, all of which is aimed at maintaining good reciprocal relationships with the ancestors. The ceremony is carried out in a joyful atmosphere and, today, Toraja people participating in it do not refrain from taking photos of themselves together with the mummified body of their dead relative (a 'ma'nene'' search on *Google* suffices to illustrate it). This confirms the strong sense of intimacy with the ancestors that Waterson noted in her own fieldwork (Waterson 2009, 207). The prayer told by the *to minaa* during the ceremony addresses the ancestors, asks them to accept the food offering, and to give blessing to their descendants in return (Nooy-Palm 1986, 203–207).

In the Sesean region, the equivalent of *ma'nene'* is called *maro*, and seems to include a further function: it contributes to completing the transformation of the dead into ancestors or divinities (*deata*) (Waterson 1993, 75). It is possible that this was also the original function of the *ma'nene'* ritual more broadly in Tana Toraja, as the expressions *ma'nene'* and *ma'tomatua* literally mean 'to make the ancestors' (Rappoport 2015, 183). The *ma'nene'* ritual can be read as a 'metamorphosis of the dead into beneficent, life-giving ancestors' (Waterson 1993, 75). Hence the importance of rewrapping or reclothing the dead bodies as a way to keep the bones together, which is essential for the dead to become (and continue to remain?) ancestors (Waterson 1993, 92–93; 2009, 207). Interestingly, rewrapping the dead is not done forever. As we observed ourselves, very old *mebalun* depositions are no longer maintained and bones end up being scattered at the bottom of the chamber, possibly becoming intermixed with others (Fig. 5.7). At which point is an ancestor 'forgotten' and no longer the object of rewrapping? This likely happens when the identity of the body is forgotten, after several generations, or perhaps when the family considers them well-established ancestors (therefore not requiring as much attention as the recently dead?).

Relocation of burials

Once placed in a *liang pa'* or a *patane*, the deceased are expected to stay there forever. Moving a corpse from one tomb to another is not part of the normal process of burial practice, and is actually rare. It only happens under particular circumstances, either because the body was placed in the 'wrong' *liang pa'*, or for sentimental reasons (for instance, moving one's close relative from a *liang pa'* to a newly built *patane* where one plans to be buried). In any case, the re-opening of the *liang pa'* and the extraction of the corpse require the sacrifice of a pig (Waterson 1995, 208–209) and up to eight buffaloes (Waterson 2009, 49).

Conclusion

The burial and ritual activities associated with *liang pa'* are very rich and diverse. They involve different actors (sponsors, relatives, ritual specialists, artisans and, in the past, slaves). From an ethnoarchaeological perspective, one can observe that these activities result in the creation of a material record at three different locations: inside the rock-cut tombs (deposition of wrapped bodies with jewellery and body-fluid jars); on the rock face immediately outside the tombs (*tau-tau* effigies, hats and cloths, vessels, food and cigarettes *etc.*); and on the ground at the foot of the rock face (*saringan* palanquins, wood vessels, bamboo poles and ladders, bones of sacrificed animals). Waste rock resulting from the cutting of the tomb, as well as remains of the stone workers' temporary quarters and animal sacrifices performed during the construction of the tomb (see Chapter 4), can be added to the latter category of material record. While bodies are hidden inside the tombs, special artefacts are displayed outside them. This demonstrates the double function of *liang pa'* tombs: first, to preserve the bodies of individuals (who have kinship links); second, to display the status of both these individuals and the prestigious ceremonies that their families have sponsored.

Notes
1. This term is more commonly used to refer to one of the rites celebrated during the second stage of high-rank funerals, and which involves the sacrifices of buffaloes at the *rante* (van der Veen 1940, 267; Koubi 1982, 92–94; Nooy-Palm 1986).
2. In the literature, the word is either spelt *saringan* (Kruyt 1924) or *sarigan* (Grubauer 1913; van der Veen 1940; Nooy-Plam 1979; 1986; Koubi 1982). Our main informant in Tana Toraja specifically pronounced it as *saringan*, therefore we have retained this spelling here. *Saringan* are also called *langi langi* in areas north of Rantepao such as Bori', or *duba duba* in the southern district of Rembon.

3   Exchange rate in June 2017: 1 GBP = 17,250 IDR; 1 USD = 13,380 IDR.
4   Volkman (1979a) and Crystal (1985) present the concept of *bombo* respectively as the spirit and the soul of the deceased. Koubi (1982, 164) has a slightly different understanding, and presents the *bombo* as a 'double' of the deceased, separate from its spirit/soul which travels to the otherworld (Puya) and is therefore no longer present after the completion of the funeral ceremony (see also Nooy-Palm 1986, 246; Waterson 2009, 377–379).
5   It is not uncommon to find *liang pa'* with commingled human remains in Tana Toraja but, in most cases, they correspond to emergency mass burials made in the years 1918–1920, when a large number of people died in Tana Toraja during the Spanish flu (Waterson 2009, 379). People buried this way did not receive traditional burial treatment (*mebalun*), and the rock-cut tombs inside which they were deposited were not properly closed and likely not consecrated, considering the emergency of the situation. These burials do not correspond to the collective burial assemblages we are dealing with in this chapter, which are recognisable by the presence of *mebalun* bodies and/or traditional wood-carved doors.

Chapter 6

The landscape setting of *liang pa'* cemeteries

In this chapter, we explore how cemeteries of rock-cut tombs interact with their surrounding landscape. The landscape considered here is both the physical setting, with particular geological and topographical properties, and the socially lived and constructed environment, embedded with cosmographic concepts and perceptions as well as with daily-round experiences. What is the role of the natural landscape in providing special places to house the dead? How does the landscape help in orchestrating spatial and social relationships between the dead and the living?

Stone and death: Toraja cemeteries as rocky places

In Chapter 2, we described how aristocratic tombs in Tana Toraja have presented various forms through time, from wooden sarcophagi to natural cliff crevices, rock-cut chambers and house-tombs. These tombs vary architecturally, but they all display an essential property, which is an immediate connection with stone and rocky places. In the past, *erong* sarcophagi were placed outside villages at dramatic cliffs or rock-shelters. Over the last centuries, rock-cut tombs, by definition, have been made into natural solid rock, and house-tombs have been made traditionally on large horizontal outcrops so that the dead can be laid out either over the flat rock surface or into hewn out vaults. To put it simply, dead nobles are always buried in the rock, never in soil. The only exception are noble infants, who are buried in trees.

The association between death and stone is not exclusive to Toraja culture but actually shared among many Austronesian cultures. From Madagascar (Parker Pearson and Ramilisonina 1998) to many Indonesian islands, such as Sumba (Hoskins 1986; Keane 1997) or Borneo (Janowski 2020), stone is used for burial monuments and the commemoration of the ancestors because of its material property of durability.

Stone is conceptually opposed to wood as an ephemeral building material dedicated to domestic structures such as houses.

In Toraja mythology, bedrock is associated with the origin place of female deities who participated in the creation of humans. The first human ancestors were created in the sky by Puang Matua (the 'Old Lord' of the sky) whose wife (Arrang di Batu, 'Light from the Stone') and mother (Simbolong Manik, 'Hair-bun like Golden Necklace-Beads') self-generated within large rock boulders and were extracted from them by their deity husbands (Waterson 2009, 129). As highlighted by Roxana Waterson, stone in Toraja mythology and traditional beliefs is associated with life-giving and fertility properties. The idea of burying the dead into natural rocks (cliffs and boulders) represents a movement back into stone that reverses the movement associated with the myths but finds its explanation in the expectation that the dead are to become deified ancestors who will become a permanent source of fertility for their descendants (Waterson 2009, 130–131). Toraja mythology and worldview, therefore, are key to understanding the important connection between the ancestors and stone (as a material associated with permanency, origin myths, and fertility), and the importance of rocky places as the focus for burial depositions.

Several remarkable rocky places in Tana Toraja were used as cemeteries over many generations, and it is not uncommon to find different traditions of burials overlapping at single sites. Ancient *erong* sarcophagi and *liang pa'* are found together at the cemeteries of Pana' (Suloara'), Parinding boulder 10, Tampangallo and Suaya (Sangalla'), as well as with *patane* at the cemeteries of Tongka' (Tallunglipu), Marante, Buntu Pune and Ke'te' Kesu. These cemeteries were used for centuries by local communities and must have held very special significance before some of them became official touristic places in recent decades.

Toraja cemeteries are created in singular rocky places and therefore they have a direct, essential connection with the natural landscape. Unlike modern, concrete *patane*, rock-cut tombs cannot be created anywhere in the landscape. However, beyond this natural determinism and prerequisite, are there any important cultural factors that influence the locational choices of tombs and cemeteries? If a given Toraja community had several rocky places in their local area that were suitable for the creation of tombs, what criteria would they first consider when making their choice? Does the orientation of the tombs matter, or the position of the cemetery in relation to the local topography and the village? We will address these questions by reviewing the available literature and referencing our own field observations.

Who owns the cemeteries? Communal *vs* private lands

The first practical aspect to consider is the status of the land on which rocky places are found in the landscape. According to the literature and our field observations, there seems to be two main models, depending on the natural configuration of the rock face and its spatial relationship to arable lands. In the first model, the cemetery is on a large cliff face that is owned collectively by a village community or a group of local *tongkonan* houses. The cliff is typically located on the edge of a wooded mountain that cannot be used for rice cultivation. The cemetery is a collective property in the same way as are the ceremonial plaza (*rante*), the village woodland (*pangala'*) and the marketplace (Nooy-Palm 1979, 92–93).

In the other model, cemeteries are located on private land. This is typically the case for most of the volcanic boulders distributed on the slope of Mount Sesean, and which are often scattered across cultivated rice paddies. The rice paddies and their boulders are owned by local *tongkonan*. Each *tongkonan* uses its boulders for its own tombs and may allow other *tongkonan* with whom it has close relationships to create their tombs there too.

The status of the land, therefore, has to be taken into consideration when exploring the landscape relationships between cemeteries and villages, and perhaps before any other (*e.g.*, cosmographic – see below) parameters. The location of cemeteries in privately-owned lands can indeed explain apparently awkward spatial patterns. The cemetery of To Bua' in Deri provides a good example (Pl. 9). The cemetery consists of a cluster of eight volcanic boulders (Deri boulders 1–8 in our database), which are immediately adjacent to the local main road, in a small woodland surrounded by several *tondok* (*tongkonan* hamlets) and rice terraces. The cemetery has been used over many generations, as indicated by the presence of two old *liang pa'* along with 14 recent ones and two newly built vacant tombs (for a plan of the cemetery see Fig. 6.11). The cemetery land belongs to a single corporate group, whose origin-house (*tongkonan* Ne' Kararo) is located 720 m (as the crow flies) to the southwest. Over generations, *tongkonan* Ne' Kararo has built close relations (typically through marriages) with another kinship group (*tongkonan* Bontong) whose origin-house is located 100 m uphill from the cemetery. Most of the tombs in the cemetery were founded by members of *tongkonan* Ne' Kararo, but a few others belong to members of *tongkonan* Bontong. As the map shows (Pl. 9), the To Bua' cemetery is not used by *tongkonan* houses located in the immediate vicinity, as one would expect, but by more distant ones. The cemetery itself belongs to (and is used primarily by) a *tongkonan* that, incidentally, is geographically closer to another cluster of tombs (Deri boulders 10–14).

The example of To Bua' shows that landscape relationships between residential areas and cemeteries can sometimes be more complex than expected. In this specific type of landscape, where tombs are made in boulders that are scattered over multiple privately-owned parcels of land, tombs do not always belong to the closest *tongkonan*.

By contrast, cliff cemeteries typically used in limestone areas of Tana Toraja generally present a simpler landscape relationship with settlements. They are collectively owned spaces, shared by a large number of *tongkonan* groups in a wider catchment. Chapter 7 will tackle the issue of personal biographic connections between specific *tongkonan* genealogies and cemeteries. Here, in this chapter, we address more general aspects of the landscape setting of cemeteries based on broadly shared cultural concepts in Tana Toraja, particularly cosmographic conceptions and perceptions of the landscape.

The importance of cardinal points in Toraja cosmography and ritual practices

Cardinal points are essential in Toraja worldview and religious life. They are not just points of reference, they actually structure the entire belief system and ritual practices (Nooy-Palm 1986; Waterson 2009, 301–302). In terms of cosmography, each cardinal point is associated with specific supernatural beings. The deities (*deata*) are associated with the east, except Puang Matua (the 'Old Lord' of the sky) who is associated with the north. The dead (*to dolo*) and their land (Puya) are believed to be somewhere far to the south or southwest of Tana Toraja. This is where the souls of the deceased travel once the complex funeral ceremonies are completed.

Ritual practices, as proscribed in the traditional religion (*aluk to dolo*, 'way of the ancestors'), include a great diversity of rites, often involving offerings and ritual speech. All of these rituals are classified into two major categories. The Rites of the East, on one hand, are addressed to the deities and are concerned with life. They include all the rites associated with house (re)building and rice cultivation. Rites of the West, on the other hand, are addressed to the ancestors and are concerned with the dead. They include all the rites associated with death, funerals and the transformation of the dead into ancestors (Waterson 2009).

Traditional Toraja cosmography determines both the *time* and *location* of ritual actions. Rites of the East must be performed in the morning (when the sun is rising), and to the east of the house (in direction of the deities), while Rites of the West must be performed in the afternoon (when the sun is setting), and to the west of the house (in direction of the ancestors). This is demonstrated in many rites associated with funeral ceremonies. During the first phases of elaborate funeral ceremonies, which take place at the village, the sacrificial chicken, pigs and buffaloes are always killed on the west side of the *tongkonan* house of the deceased individual for whom the ceremony is held (Kruyt 1924, 139, 160; Koubi 1982, 60–160; Nooy-Palm 1986, 175). Before the start of the funeral ceremony, the dead person is considered 'sick' and is laid out in an east–west configuration inside the *tongkonan*, which is the way living people lie down to sleep. Once the funeral ceremony begins, the deceased is officially considered 'dead': at this point, the body is moved from the back (sleeping) room of the house to the middle room of the house and is laid out north–south, with the head to the south, as a symbolic way to help the deceased in his/her departure to Puya[1] (Kruyt 1924, 139; Koubi 1982, 158; Nooy-Palm 1986, 195, 227–228; Waterson 2009, 380–381). During the period prior to the departure of the body from the house, the widow(er) sleeps next to the deceased, also with their head turned to the south (Kruyt 1924, 153; Nooy-Palm 1986, 185, 229). Finally, once all the funeral is completed, and the dead placed in the grave, the widow(er)'s mourning and rice prohibition is ended by a rite consisting of placing an offering of boiled rice on the ground on the west side of the house (Kruyt 1924, 156), or cutting the widow's hair while she sits on the west side of the house, facing west (Nooy Palm 1986, 177–178). The west side of the house is also the location where stillborn children are buried (Kruyt 1924, 138),[2] while the placentae of living newborn babies are buried on the east side of the house (Waterson 2009, 185).

Cardinal directions, therefore, are crucial in both ideal conceptions (cosmography) and material practices of ritual activities associated with life and death. They also determine the way traditional villages are structured and orientated, with rows of *tongkonan* houses facing the north, and *alang* rice barns placed in front of them and opening to the south. Are cemeteries the object of similar rules? Are tomb orientations and landscape locations of cemeteries determined by specific cardinal directions?

Landscape relationships between cemeteries and villages

According to ethnographic literature, cardinal directions and natural topography were traditionally taken into consideration when choosing the landscape location of tombs so that the relative position of villages and cemeteries reflected principles of Toraja cosmography. According to Jeannine Koubi, in both Sa'dan and Mamasa regions, a cemetery was always located to the south or southwest of its corresponding village by a distance of up to more than 1 km (Koubi 1982, 195, 229, 272). Koubi also claims that the ceremonial plaza (*rante*) is located to the south, west or southwest of the village so that the funeral procession from the village to the plaza symbolically imitates the journey of the dead toward Puya (Koubi 1982, 180). Unfortunately, Koubi does not make reference to specific villages or cemeteries to support these claims.

More recently, Akin Duli also claimed that traditional customs dictated that cemeteries be located to the southwest of villages (or their founding *tongkonan* houses), and on a higher landscape position so that the ancestors could permanently monitor the activities of the living (Duli 2018, 47–48, 52; Duli *et al.* 2019, 7–8). Duli cites the example of the villages of Ke'te' Kesu (Kesu), Bori' (Sesean), Pallawa' (Sesean), and Sillanan (Gandang Batu Sillanan) to illustrate this model. Although the model seems to apply to the latter two villages, it does not for the former two. At Bori', the cemetery complex is located to the north of the ancient village,[3] not to the south (see Pl. 13). At Ke'te' Kesu, the location of the founding *tongkonan* (*tongkonan* Kesu') changed over time. According to Duli (2018, 42–44), it was originally located 1 km to the southeast of the current location of the village (Pl. 10) and was therefore located to the southeast of the cemetery too.

In reality it seems very unlikely that the location of cemeteries in the past was decided on the basis of cardinal directions and relative landscape position with the village. Cemetery locations were more likely decided on the basis of multiple practical factors (see below) rather than on intentions to apply a general cosmographic model to the reality of the physical landscape. The latter presents its own constraints and opportunities which vary from one village area to another and were used opportunistically on a case-by-case basis.

From a methodological point of view, it would even be quite difficult to determine preferred cardinal directions in cemetery–village relationships. First, cemeteries are typically used by several *tongkonan* distributed in various locations across the landscape. Since old traditional cemeteries have been used over centuries it is often (not always) difficult to determine which were the first *tongkonan* to use them, and where they were located in the landscape. Second, village locations have often moved since the time their *tongkonan* and associated tombs were founded. This has consequences on spatial relationships as one can observe them today. Ancient villages were often built on hilltops for security reasons and were protected by walls (Nooy-Palm 1979, 12, 250). The arrival of the Dutch in the early 20th century put an end to inter-village warfare and headhunting. As a result, several ancient villages were progressively relocated closer to the plains, rice paddies and main roads. Ke'te' Kesu, discussed above, is one example. Another is Buntu Pune, 1.75 km to the northeast of Ke'te' Kesu, which is one of the oldest villages in Tana Toraja (Pl. 11). The current village (*tondok*) of Buntu Pune is located on a terrace created on the northern slope of Buntu Kongkang Mountain. It consists of two old *tongkonan* houses and four *alang* rice barns with beautiful traditional bamboo thatch roofs. From the village, a small path leads to the foot of a cliff *c.* 100 m to the west, with old *erong* sarcophagi, a few *liang pa'* and many *patane* tombs. Another path leads from the village to a rock plateau located on the top of the cemetery cliff, which corresponds to the location of the original village of Buntu Pune, of which only ruins of the old stone fortified walls and internal retaining walls can be seen today. According to a local resident, the village of Buntu Pune was founded on the high plateau and existed for ten generations before being relocated in *c.* 1880. Lack of space on the plateau and the decrease in violence at the end of the 19th century convinced the community to relocate the village down to its current location.[4] The moving of Buntu Pune from the hilltop plateau to its current location has affected its cardinal relationship with the cliff cemetery: while the old village was located to the south of the cemetery, the current one is now located to the east. The latter configuration may be mistakenly regarded as a deliberate reproduction of the traditional cosmographic model while, in reality, it is just an unintended coincidence resulting from a local historical process.

In the southwest district of Rembon, where we could accurately map specific *tongkonan* associated with local cemeteries (see Chapter 7), all villages were located to the south of the cemeteries, which again goes against the cosmographic model. When we asked *to minaa* Paulus Kondosara about cardinal directions, he said these are important in ritual practices, but are not at all considered when deciding on the location of a tomb. People, even in the past, could create tombs in rock faces located at any direction from their *tongkonan*.

Cardinal points and cosmography, therefore, do not seem to have played any role in landscape relationships between villages and cemeteries. Our field enquiry reveals, instead, that other criteria were considered, in particular, the distance between houses and tombs and the visibility of the tombs. In the past, cemeteries were considered sacred places and were only visited on rare occasions. Cemeteries were often located in discrete, remote locations, which would guarantee their safety (*e.g.,* they could not be easily found by raiding troops) and would ensure that the dead would not be disturbed. Several old cemetery cliffs such as Londa, Pana' and Lombok Parinding, are located within small landscape depressions that are invisible from surrounding villages. They consist of dead-end places that are located outside of road networks. Many other old cemetery cliffs or boulders are well hidden in dense forests or distant hill slopes that are only accessible by small paths. This contrasts with recent liang pa' and patane tombs, which are often made on the side of roads and are intended to be seen by all passers-by, sometimes displaying conspicuous carvings.

Several informants explained how, especially in the past, people, and particularly children, were afraid to go to the ancient, secluded cemeteries. This was also reported by Waterson, who adds that strange occurrences were common experiences at these sites (Waterson 2009, 365).[5] It is only with the development of the Christian religion in the Toraja territory that spatial distances between people and the dead have been reduced, as represented by concrete patane tombs that are often built close to villages or anywhere conveniently accessible. The issue of the distance and visibility of tombs is further discussed below in our next section dealing with the landscape orientation of liang pa' tombs.

Landscape orientation of the tombs

In all Toraja villages, the orientation of domestic structures systematically follows the same pattern, with rows of tongkonan houses facing north (sunlight), and rice barns placed a few metres away in front of them to the north. In the Mamasa region, traditional batutu house-tombs (and sometimes even the old sarcophagi they contain) are orientated to the south, in the direction of Puya, the land of the dead (Duli 2014). In the Sa'dan region, we observed an ancient wooden patane in Tembamba which was also orientated to the south, with its small opening door on the southern wall of the structure (Chapter 2). Do liang pa' tombs have to follow similar proscriptions? According to Grubauer, who visited Tana Toraja in 1912, the cardinal orientation of tombs had no importance in the areas he explored: some liang pa' opened to the southeast, while others were directed to the west or the northwest (Grubauer 1913, 214). Our systematic survey of 697 liang pa' from 17 sites and areas gives us an opportunity to evaluate statistically whether orientations were meaningful with Toraja rock-cut tombs. In the present section, we explore general patterns. In the section that follows, we offer a more detailed, site-specific approach.

The orientation of a rock-cut tomb is dependent on the orientation of the rock face in which it is created. That is why it is important to make a distinction between cliff cemeteries (where all the tombs adopt the same orientation imposed by the face of the cliff), and boulder cemeteries (where tombs can be made on different faces of one or several boulders and, therefore, may present multiple orientations locally). One could argue that boulder cemeteries, theoretically, offer a wider range of options in terms of tomb orientation.

Table 6.1 presents the main orientations of *liang pa'* in cliffs and boulders from our survey database. This first analysis shows that, overall, there is no significant patterning with regard to tomb orientation. Statistically, tombs are more often orientated to the northeast (23%), southwest (15%), southeast (14%), south (13%) and east (13%), while they are less often orientated to the northwest (9%), west (8%) and north (5%). The dominant northeast orientation does not support the idea that tomb orientation was based on the directional cosmographic principles associated with the dead, those being the south (which is the orientation of *patane* tombs and bodies during funeral ceremonies) and the west (direction of offerings addressed to the dead).

Table 6.1. Orientations of the 697 liang pa' rock-cut tombs surveyed in June 2017, in relation to their landform context (cliffs vs boulders).

Orientation	Cliff tombs	Boulder tombs	Total
North	8	26	34
Northeast	83	78	161
East	9	81	90
Southeast	9	86	95
South	14	80	94
Southwest	0	103	103
West	32	27	59
Northwest	19	42	61
Total	174	523	697

Is the dominant northeast orientation of rock-cut tombs otherwise meaningful culturally? The front sides of *tongkonan* houses theoretically face north but, in practice, they tend to face northeast (*i.e.*, the morning light) – as can be easily verified by looking at satellite photographs of the region (*e.g.*, Pl. 9). Is there an intention for the houses of the dead to replicate the orientation of the houses of the living? *Liang pa'* tombs are explicitly referred to as houses of the dead in ritual poetry (*tongkonan tangmerambu*, 'the houses without smoke/where no fire is lit'). This is a tantalising hypothesis which, unfortunately, is hard to support further. The orientation of tombs generally does not seem to hold any specific meaning in traditional funerary practices and, statistically, the dominant northeast tomb orientation is not very meaningful, being represented by less than a quarter of *liang pa'*.

On the contrary, tomb orientations seem to be determined by local terrain conditions (see below) rather than specific cosmographic proscriptions. We can see this when comparing the orientations of cliff tombs with boulder tombs. One can notice slight differences between cliff and boulder tombs (Fig. 6.1). The majority of cliff sites are exposed to the northeast, while boulder sites show more frequent orientations to the southwest, southeast and east, representing a much more diverse range of orientations than cliff tombs.

Let us now look at patterns through time, by comparing the orientations of old (1700s–1960s) and recent *liang pa'* (1960s–today) (Table 6.2; Fig. 6.1). No major, meaningful differences can be noted overall between the general orientation of old *liang pa'* and that of recent ones. Recent *liang pa'* are not represented by a dominant orientation, with the southeast, southwest and northeast being the most common

6. The landscape setting of liang pa' cemeteries

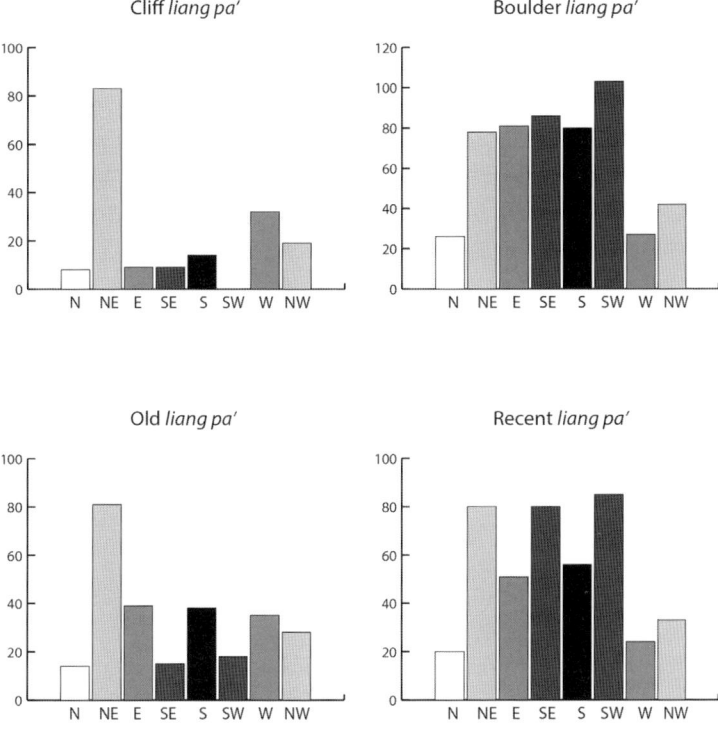

Figure 6.1: Orientations of the 697 liang pa' rock-cut tombs surveyed in June 2017, in relation to their landform context (cliffs vs boulders – top) and date (old vs recent – bottom) (graphs: G. Robin).

orientations by nearly equal measures. The majority of older tombs, by contrast, face northeast. These patterns are consistent with the one observed at cliff vs boulder sites and are explained by the fact that most recent tombs (400/443, 90%) are created in boulders, while old liang pa' are more often found on cliffs (145/254, 57%).

A last aspect to consider is local variability in tomb orientation, which may be influenced by different landscape configurations. Table 6.3 presents tomb orientations by individual cemeteries or localities: the first seven rows in the table are cliff cemeteries, with rock face orientation determined by large scale geological faults; the subsequent

Table 6.2. Orientations of the 697 liang pa' rock-cut tombs surveyed in June 2017, in relation to their date (old vs recent).

Orientation	Old liang pa'	Recent liang pa'	Total
North	14	20	34
Northeast	81	80	161
East	39	51	90
Southeast	15	80	95
South	38	56	94
Southwest	18	85	103
West	35	24	59
Northwest	28	33	61
Total	268	429	697

Table 6.3. Orientations of the 697 liang pa' rock-cut tombs surveyed in June 2017, in relation to slope aspect of their landscape setting.

Locations		Slope aspect	No. tombs	N	NE	E	SE	S	SW	W	NW
Batu Lappa'	Cliff	–	14								14
Buntu Pune	Cliff	–	1	1							
Ke'te' Kesu	Cliff	–	4								4
La'bo'	Cliff	–	3						2	1	
Lemo	Cliff	–	80			80					
Suaya	Cliff	–	6				5	1			
Tampangallo	Cliff	–	10	6			4				
Batutumonga	Boulders	SW, SE	47	0	5	2	16	5	11	1	7
Bori'	Boulders	SE	102	5	6	19	15	26	19	5	7
Buntu Lobo	Boulders	S, SE	58	6	0	5	16	18	6	5	2
Deri	Boulders	SE	31	1	7	7	3	1	8	1	3
Lempo	Boulders	SE	72	10	10	15	8	6	12	5	6
Marante	Cliff	–	3	0	3	0	0	0	0	0	0
Parinding	Boulders	SE	62	5	5	5	13	13	8	3	10
Tonga Riu	Boulders	SW	100	0	12	26	20	5	32	5	0
Suloara	B./Cliff	S, SW	93	0	33	2	4	6	7	33	8
Tallunglipu	Cliff	–	11	0	0	0	0	11	0	0	0
			697	34	161	90	95	94	103	59	61

ten rows are local groups of basaltic boulders located on the southern slopes of Mount Sesean, which present various faces and therefore offer multiple choices in terms of tomb orientations. For the boulders, the dominating slope aspect is indicated by shading for each local group.

The main outcome of this analysis is a clear statistical correlation between slope aspect and the orientation of the tombs in local groups of boulders. The most frequent tomb orientations correspond to the direction of the slope within these groups. This can be explained by how these boulders present themselves in local terrain conditions. Boulders selected to receive rock-cut tombs are often positioned half buried into the hill slopes; therefore, the most exposed vertical faces of the boulders are generally the ones facing down the slope, while the most buried examples face upslope. As a general rule (to which there are exceptions of course), people prefer to use the widest rock face to create tombs, especially since most boulders in the Sesean region are only big enough to receive one or two tombs. Based on our survey, the average number of tombs per boulder is three (523 tombs for 173 boulders). If we exclude

Table 6.4. Proportions of boulders from Mount Sesean (study area A) having liang pa' rock-cut tombs in one or more faces.

	Tombs on 1 face	Tombs on 2 faces	Tombs on 3 faces	Tombs on 4 faces	Tombs on 5 faces	Tombs on 6 faces
No. boulders	126	30	9	3	1	4
% total	73%	17%	5%	2%	1%	2%

exceptionally large boulders like Lo'ko' Mata in Tonga Riu (95 tombs) and Sele in Suloara' (30 tombs), the average is 2.3. Based on these figures, one may conclude that boulders in the Sesean region are often too small to receive multiple tombs and, in individual boulders, tombs will normally be cut in the wider face, which looks down the slope. Therefore, in the same way that cliff tombs simply adopt the orientation of the natural cliff, boulder tombs tend to adopt the natural orientation of the slope. All of this supports the idea that in Tana Toraja, overall, the orientation of rock-cut tombs has been mainly determined by practical terrain configurations rather than cosmographic principles.

If we want to explore cultural choices beyond natural determinism, one needs to identify boulders having tombs on more than one face. Table 6.4 shows that 27% of boulders in Mount Sesean (study area A) have tombs on multiple faces (*i.e.*, presenting different orientations).

Why were tombs made on different faces on these boulders? Was it for purely pragmatic reasons, *e.g.*, the first face was running out of space and therefore later tombs were made on the other faces available? If so, why, on these boulders, were specific faces preferred over the other ones in the first place? Can we identify patterns in the way rock faces were selected on boulders, based on their morphology, orientation, location or visibility from a distance? Have selection criteria changed through time?

Boulder cemeteries: case studies

The best way to address the questions above is to focus on selected case studies of cemeteries that include boulders with both old and recent *liang pa'*, so that we can observe how the location and distribution of tombs have developed through time within the micro-landscape of the cemetery.

Lo'ko' Mata (Tonga Riu boulder 1)

The largest boulder we encountered during our survey is the famous rock cemetery known as Lo'ko' Mata (Tonga Riu boulder 1 in our dataset). The cemetery's name derives from the Toraja words 'hole' (*lo'ko'*) and 'eyes' (*mata*). This huge boulder (c. 30 × 40 m wide, 20 m high) indeed looks like a supernatural rock pierced with dozens of eyes (Figs 2.13; 6.2). Although it is geologically a boulder, it can be considered a small mountain with multiple cliff sides. The boulder has a total of 95 rock-cut tombs, which are distributed all around the rock, except on its northern face. The absence of tombs there might be explained by the irregular surface of the rock face on this part of the boulder.

The 95 tombs include 13 old *liang pa'*, 77 created in recent decades, three *liang pa'* which, during our visit in June 2017, had just been completed but were still vacant and two that were in the process of being cut. Despite this impressive number of tombs, the rock still has spaces for more and we noted five reserved locations on its faces. The tombs present a variety of styles and decorations, ranging from simple ones with plain (undecorated) wooden doors, to classical *pa'tedong*-decorated tombs, and more conspicuous ones with colour-painted entrances, rock-cut balconies with *tau-tau* effigies and rock-sculpted buffalo heads (*kabongo'*). The cemetery has been the focus of funerary activity for generations and provides an interesting case study to start with.

Table 6.5 shows how older and more recent *liang pa'* are distributed on the multiple faces of the boulder. At Lo'ko' Mata, old tombs were made on the eastern and southwestern faces only, while more recent ones were created virtually everywhere around the boulder. On the eastern and southwestern faces of the rock, older tombs tend to be located at the highest positions, *i.e.*, the most difficult ones to reach with bamboo ladders. Recent tombs, on the other hand, were

Figure 6.2: Lo'ko' Mata boulder in Tonga Riu (Sesean Suloara'), with its 95 rock-cut tombs distributed on multiple faces (photos: G. Robin).

cut in more varied locations, including many that were created at ground level, making them easily accessible without the need for a bamboo ladder.

Finally, the different faces of the boulder present different orientations in relation to the main road that passes in front of the boulder, which means that certain tombs are in discreet locations, while others are well exposed to public view and all passers-by. The southeast face, in particular, is directly facing the road as one comes up to the site from the valley and enters the *lembang* of Tonga Riu (at the border of which lies Lo'ko'

Table 6.5. Orientation of old and recent liang pa' *on the boulder cemetery of Lo'ko' Mata in Tonga Riu (Sesean Suloara').*

Lo'ko' Mata (Tonga Riu boulder 1)		
Boulder faces	Old liang pa'	Recent liang pa'
NE	0	11
E	7	18
SE	0	19
S	0	3
SW	6	26
W	0	5
NW	0	0
Total	13	82

Mata). It is therefore not surprising to see that the tombs with the most conspicuous ornaments (paintings, sculpted *kabongo'*, exposed *tau-tau*), and therefore most concerned with display, are located precisely on the southeast face of the boulder.

Parinding

The *lembang* of Parinding comprises 16 *liang pa'* boulders, as well as a remarkable secluded rock shelter (Lombok Parinding) with dozens of abandoned *erong* sarcophagi (Pl. 12; see also Chapter 2). The boulders are grouped together into two main geographical clusters: boulders 1–4 in the west, and boulders 5–16 in the east. Each cluster includes both old and recent *liang pa'*. Each cluster might have constituted two separate cemeteries, which were used by different local communities over time.

The western cluster is located 250 m north of the old *erong* cemetery of Lombok Parinding. The former may have 'taken over' from the latter in the 18th–19th centuries when the custom of *erong* burials became progressively replaced by rock-cut tombs. The boulder cluster lies in a forest within a sparsely populated area. It consists of only four boulders, representing 20 tombs. Twelve of these tombs are old *liang pa'*, which are all concentrated in boulder 4 (the largest of the four). The other eight tombs are all more recent and are still being used today.

The eastern cluster is also in a woodland, at the northern margin of a densely populated area alongside the road between Rantepao and Bori', and rice paddies. It has 12 boulders of varying sizes, representing a total of 42 tombs. Boulder 10 lies at the centre of the cluster. It presents a dramatic position (resting on top of a small limestone outcrop) and an unusual flat morphology, which gives it the appearance of a large floating pebble (Fig. 6.3). Boulder 10 is certainly the oldest burial place of the cluster. It first served as a shelter for *erong* sarcophagi (four of which are still visible

Figure 6.3: Boulder 10 in Parinding (Sesean). This large, pedestalled erratic was used as a rock-shelter for erong sarcophagi and subsequently hewn out with six liang pa' *rock-cut tombs (photo: G. Robin).*

under the western side of the boulder) and was then hewn out with six *liang pa'* (on its northwest, southwest and southeast sides), which all belong to the old generation. Accessing these old tombs would have required the use of bamboo ladders. Boulder 10, therefore, can be regarded as the origin point of the cemetery. The cemetery subsequently expanded to other boulders within an initial inner circle around boulder 10 (boulders 7, 9, 11, 13, which have both old and recent tombs) and then further afield with boulders 5, 6, 8, 12, 14, 15, and 16 (which only have recent tombs).

An examination of boulders that have both old and recent *liang pa'* enables us to identify patterns in how rock faces were selected in the past. Boulder 4 is the largest in the western cluster (*c.* 15 × 15 × 15 m). It is cuboid, with four main vertical faces, all occupied by rock-cut tombs. There are 17 *liang pa'* in the boulder: 12 old and five recent ones. The distribution of tombs differs according to age, as shown in Table 6.6. Interestingly, the older *liang pa'* are all concentrated on the southern and western faces of the boulder, avoiding the northern and southeastern faces, which have been

6. *The landscape setting of* liang pa' *cemeteries* 135

Figure 6.4: Boulder 7 in Parinding (Sesean) (photo: G. Robin).

used only recently for newer tombs. How can this pattern be explained? The northern and southeastern faces are lower than the other two, and present irregular, convex surfaces. The first users of the boulder may have preferred to use exclusively the taller and flatter vertical faces of the boulder, which they found on the southern and western sides. It is possible, however, that the orientations of the faces played a role: the south and west directions being culturally associated with the dead. The rock faces with these orientations may have been considered even more desirable in the past when they were used for the first rock tombs in the area.

Moving to the eastern cluster of Parinding, three boulders have both old and recent *liang pa'*. In boulder 7, the old tomb is on the northwest face, and the recent one on the southwest (Fig. 6.4). Reasons for preferring the northwest face in the past over the southwest one may include the following: it is taller (offering higher and safer position), it is more discreet (not facing the path and plaza from which one can access the boulder), and it is slightly sub-vertical, therefore providing more shelter against the rain.

In boulder 11, the old tomb is on the east face, and the recent one is on the southwest face. The difference between the two faces is not striking, but the eastern one is slightly taller, which makes access to the tomb more difficult, requiring a ladder. Moreover, unlike the western face, the eastern face presents a small protruding cliff top which creates a convenient natural shelter for the entrance of the tomb.

Boulder 9 has 13 *liang pa'*, including five old and eight recent ones (Table 6.7). Two concrete *patane* were also built on top of the boulder. As is the case for boulder 4, old and recent *liang pa'* are distributed somewhat differently: the former are concentrated on the eastern, southeastern and southern faces of the boulder, while the latter are distributed on the south, southwest, northwest and northeast (Fig. 6.5).

Table 6.6. Orientation of old and recent liang pa' *on Parinding boulder 4 (Sesean).*

Boulder faces	Parinding boulder 4	
	Old liang pa'	Recent liang pa'
N	0	2
SE	0	3
S	9	0
W	3	0

Table 6.7. Orientation of old and recent liang pa' *on Parinding boulder 9 (Sesean).*

Boulder faces	Parinding boulder 9	
	Old liang pa'	Recent liang pa'
NE	0	1
E	2	0
SE	2	0
S	1	1
SW	0	2
NW	0	4

Here, the reasons for selecting the east-to-south half of the boulder in the past seem rather clear: the faces are subvertical, providing a natural shelter to the entrance of the tombs and to offerings deposited at the foot of the wall. The other faces of the boulder present a different morphology: they are slightly sloping, therefore more exposed to rain and running water. Recent tombs, with their deep entrance recesses and rocky thresholds under the wooden door, are better adapted to rainfall, which explains why sloping faces were more recently considered appropriate for tombs in this boulder and elsewhere in Tana Toraja. Old tombs, with their simpler entrance design (see Chapter 3), were more vulnerable to rainfall and, for this reason, were created on sub-vertical, *i.e.*, sheltered, rock faces. An additional motivation for selecting the east-to-south faces at Parinding boulder 9 is their discreet location (hidden from the path) and their taller dimension, making the tomb entrances difficult to access without the use of a ladder. Recent tombs, by contrast, are more concerned with accessibility and visibility, and are located on faces, often at ground level, more convenient for these purposes.

Bori'

The *kelurahan* of Bori' is famous for its *rante* with tall menhirs and is one of the main visitor attractions in Tana Toraja today. The *rante* is located on the northeast edge of

6. The landscape setting of liang pa' cemeteries

Figure 6.5: Plan and section of boulder 9 in Parinding (Sesean) showing locations of old and recent liang pa' rock-cut tombs (photos and drawing: G. Robin).

the old village of Bori', close to the main road and rice paddies in the plain (Pl. 13). From the *rante*, one can walk up over to a forested slope that hosts the rock-cut tomb cemetery. The cemetery is one of the most important in the region, with 102 *liang pa'* in 28 boulders concentrated within *c.* 2.7 hectares. During our 2017 survey, we recorded 43 old *liang pa'* distributed in ten boulders, 51 recent tombs in use, four tombs recently completed and vacant and four still being cut at the time of our investigation. As in Parinding, the oldest tombs were concentrated on the largest and tallest boulders (nos 1, 4 and 21). The use of smaller boulders began only later, when tombs of the recent generation came into use. Here we will look at four boulders that have both old and recent *liang pa'* distributed on several faces.

Boulder 1 is the largest boulder in Bori' cemetery. It has 26 rock-cut tombs, 17 old ones and nine recent ones. The majority of the old tombs are concentrated on the east, north and southeast faces of the boulder, while recent ones are mainly in the

Figure 6.6: Boulder 1 in Bori' (Sesean) seen from east (left) and southwest (right) (photos: G. Robin).

southwest, west and northwest faces (Fig. 6.6; Table 6.8). This distribution is explained by the morphology of the boulder and by the selection criteria identified in the Parinding example. The faces with older tombs are the tallest ones (allowing high, secure positions to the tombs), and their walls are sub-vertical (thus less exposed to rains). Recent tombs, on the other hand, are made on faces that are shorter and sloping. The only recent tomb made on the eastern face of the boulder (face normally associated with old tombs), is positioned at ground level.

Table 6.8. Orientation of old and recent liang pa' *on Bori' boulder 1 (Sesean).*

	Bori' boulder 1	
Boulder faces	Old liang pa'	Recent liang pa'
N	3	1
E	11	1
SE	1	0
SW	2	3
W	0	2
NW	0	2

Boulder 4 presents exactly the same principles (Fig. 6.7). The seven old *liang pa'* are distributed on the taller, sub-vertical faces of the boulder (south, southwest, northwest and northeast) and are high above the ground level, necessitating a ladder to be accessed. The only recent tomb in the boulder was cut in the only short, sloping face available (southeast side), and is close to the ground.

At boulder 10, two recent and conspicuously decorated *liang pa'* were created at ground level, facing southeast. The tombs have a unique painted and sculpted ornamentation and they were made on the shorter, slightly sloping face of the boulder. By contrast, the single old *liang pa'* was created high up at the summit of the northeast face of the boulder, which is the narrowest but tallest face of the boulder, and also the most hidden. Indeed, the face is located on the opposite side of the path going up the hill. The northeast face is also partly obscured by another boulder located in front of it. Boulder 10 shows very well how old and recent *liang pa'* have completely opposite strategies of access and visibility.

6. The landscape setting of liang pa' cemeteries

Boulder 21 is the second largest boulder in Bori' and contains 24 rock-cut tombs (nine old and 15 recent) (Fig. 6.8; Table 6.9). Most of the old *liang pa'* are on the southern face of the boulder, although two are on the west-northwest faces. Were these faces chosen for the orientations, cosmographically associated with the dead and the ancestors? This possibility cannot be excluded, but it was probably their physical properties that most-of-all attracted the first tomb carvers, since they are sub-vertical and the tallest faces of the boulders. All the old *liang pa'* at boulder 21 are located high up on the rock faces and cannot be reached without a ladder. Recent tombs, on the other hand, have used the remaining spaces available, i.e., either short, sloping faces, or lower positions on previously used tall faces.

Table 6.9. Orientation of old and recent *liang pa'* on Bori' boulder 21 (Sesean).

Boulder faces	Bori' boulder 21	
	Old *liang pa'*	Recent *liang pa'*
NE	0	4
SE	0	2
S	7	4
W	1	2
NW	1	3

Bori' is a very rich and interesting *liang pa'* site. It is a large, well-bounded cemetery with multiple boulders, some of them having both old and new tombs. Each of these boulders present a different morphology and terrain configuration. They have rock faces with different orientations and heights, and they present different levels of visibility and accessibility. All of these parameters were very clearly taken into consideration by Toraja tomb cutters, both in the past and in recent times. By comparing the location of old and recent tombs we have determined that the same patterns applied in both Bori' and Parinding. Stone cutters in the past had a clear set of criteria to select suitable rock faces: the latter had to be tall enough to accommodate unreachable tomb entrances; they had to be sub-vertical, with a protruding top in order to give as much shelter as possible to tomb entrances; finally, where possible, they had to be facing away from main paths and open spaces so as not to attract the attention of looters. If cosmographic orientations were taken into consideration, they probably represented only secondary criteria. In more recent times, with declining levels of insecurity (e.g., tomb raiding; Keers 1939, 207), and the increase in conspicuous funeral expenditures and social display, tomb cutters have adopted another set of criteria, indeed a flipped pattern compared to the previous generation: they have prioritised easily accessible positions on rock faces (not requiring ladders) and high visibility landscape locations to be viewed by as many people as possible. With the development of cutting techniques and expenses, recent tombs have adopted structural solutions to protect their contents from rainfall and have consequently been less concerned with finding sheltered rock faces.

So far, we have focused on the relationships between the tombs and the physical configuration of the boulders. Our last two case studies revisit another issue touched upon above: the visibility of the tombs and how their distance from settled areas has shifted through time.

Figure 6.7: Boulder 4 in Bori' (Sesean) seen from southwest (top) and southeast (bottom) (photos: G. Robin).

Buntu Lobo (boulders 9–13)

Buntu Lobo is a *lembang* with a particular abundance of basalt boulders and outcrops. During our survey of the area, we noted 33 boulders (representing 58 rock-cut tombs), as well as several active quarries for menhirs and sculpted stone bases of *tulak somba* support pillars for *tongkonan* houses. As in other areas of the Sesean highlands, boulders with tombs are grouped together in various localised clusters. The cluster with

6. The landscape setting of liang pa' cemeteries

Figure 6.8: Boulder 21 in Bori' (Sesean) seen from the south. Old liang pa' rock-cut tombs are indicated with an asterisk (photo: G. Robin).

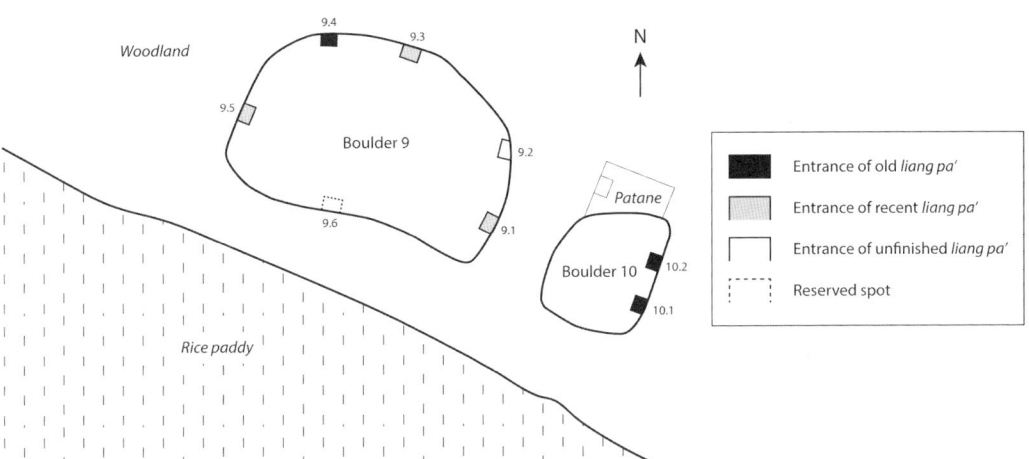

Figure 6.9: Plan of boulders 9 and 10 in Buntu Lobo (Sesean) showing locations of old and recent liang pa' rock-cut tombs (drawing: G. Robin).

boulders 6–16 is of particular interest, as it includes both old and recent *liang pa'* in the same landscape composed of woodlands, rice paddies and settlements (Pl. 14). The oldest tombs were made in boulders 9 and 10, which are located 100 m away from the nearest houses, at the boundary between a woodland (to the north) and rice paddies (to the south) (Fig. 6.9). Boulder 9 contains five tombs (including one that is old) and boulder 10 contains three tombs (including two that are old). Old tombs were made on the northern and eastern faces of the boulders, *i.e.*, facing the woods and away from the closest settlement. More recent tombs were created in boulders located around the historical 'core' (boulders 9–10), but closer to roads and houses. Apart from boulders 11–13, those with recent tombs are well exposed to views from the main road across the area. According to our guide, Amos Palungan, many tombs are cut into boulders that are adjacent to roads for practical reasons of accessibility, but also so the tombs are more visible on a daily basis and the deceased are better remembered.

Suloara' (boulders 3–7)

Our last case study is located in the *lembang* of Suloara'. It is a cluster of five boulders concentrated on a wooded slope on the side of the *lembang*'s main road (Fig. 6.10). The cemetery also includes two concrete *patane* standing side-by-side immediately adjacent to the road, as well as a flat, ground burial of a Muslim individual. The five boulders include old and recent rock-cut tombs, whose locations enable us to reconstruct the spatial development of the cemetery through time. The oldest *liang pa'* are located in boulders 6 and 7, which are the highest in the slope and most distant from the road. In a second phase, recent style *liang pa'* were built in boulders 3 and 5. These are closest to the road and the tombs themselves face the road and are visible from it (unlike boulders 6 and 7, which are hidden in the woods). Tombs in boulders 3 and 5 were probably built around the same time as the two concrete *patane* on the edge of the road. In the third phase, three tombs were built in the only remaining places in the cluster, *i.e.*, in boulders 4 and 6. The single tomb in boulder 4 was still in the progress of being cut during our visit in June 2017, while the two new tombs in boulder 6 had just been completed and were still vacant.

The case study from Suloara' exemplifies how a boulder cemetery develops through time, and how shifting criteria for the location and visibility of tombs over old and recent tomb generations affect the spatial layout of tombs within a boulder cemetery. A similar development can be noted at the cemetery of To Bua' (Deri boulders 1–8) in Sesean (Fig. 6.11).

Conclusion

In this chapter we have explored how rock-cut tombs interact with their physical, natural environment (geology, topography), as well as with human taskscapes (*sensu* Ingold 1993: villages, rice paddies, woodlands). Our analysis was carried out at three different scales: the rock (the morphology of boulders and rock faces and how they influence the placement of tombs); the cemetery (how tombs are distributed within groups of boulders, based on their age, orientation and decoration); and the landscape

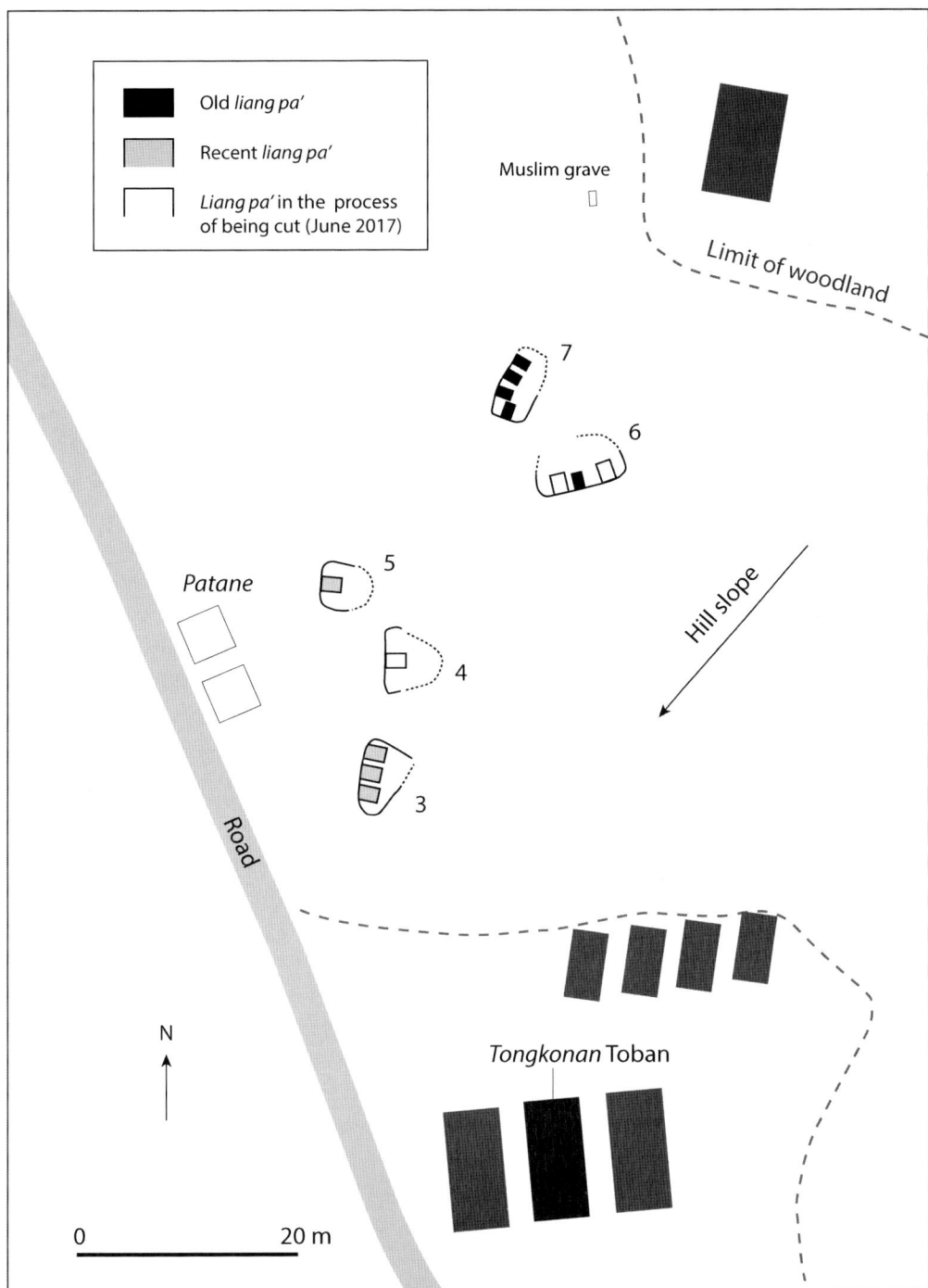

Figure 6.10: Map of boulders 3-7 and surroundings in Suloara' (Sesean Suloara') showing locations of old and recent liang pa' *rock-cut tombs as well as those which were in the process of being cut at the time of our investigation (drawing: G. Robin).*

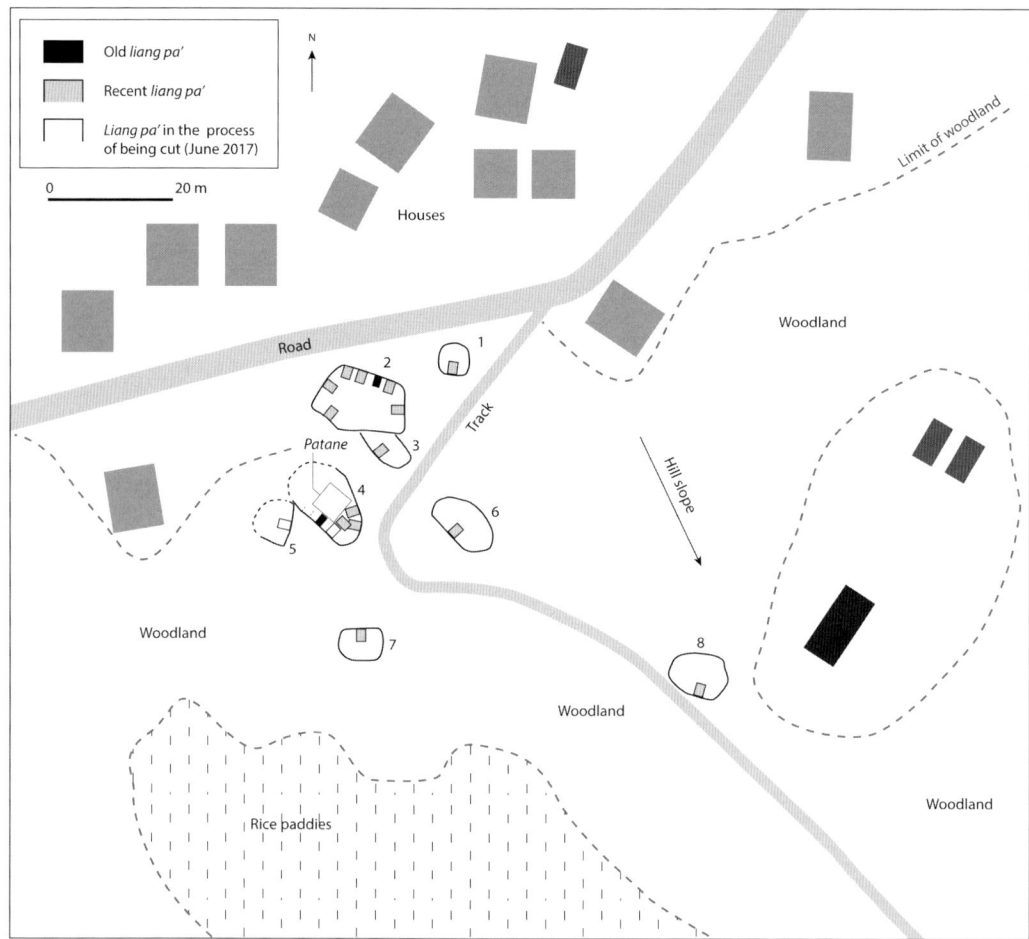

Figure 6.11: Map of boulders 1–8 and surroundings in Deri (Sesean) showing locations of old and recent liang pa' rock-cut tombs as well as those which were in the process of being cut at the time of our investigation (drawing: G. Robin).

(how boulders and cemeteries relate spatially to residential areas and cultivated land parcels).

The starting point of our enquiry was the landscape orientation of the tombs. Toraja mythology, cosmology, ritual practices and domestic architecture are strictly orientated in the landscape in very specific ways, which have been systematically reproduced over generations throughout all of Tana Toraja. In this broad cultural model, the south and the west are exclusively associated with the dead and the ancestors, and one could expect that tombs, as formal ritual structures, would take these orientations into consideration. Our study shows that this is not the case. Unlike what has been claimed in the literature, it is unlikely that the cardinal orientation of rock faces and their topographic relation to villages, played any role in the process of

6. *The landscape setting of* liang pa' *cemeteries*

Table 6.10. Main differences in the landscape location of old and recent liang pa' *rock-cut tombs.*

Criteria	Old liang pa'	Recent liang pa'
Type of rock face	Tall, flat, overhanging, sheltered	Any, including exposed sloping faces
Tomb position	High up, requiring ladder for access	Close to ground level, no ladder needed
Location	Away from paths/roads and villages	Close to paths/roads and villages
Orientation	Away from paths/roads and villages	Towards paths/roads and villages
Public visibility	Minimum	Maximum
Accessibility	Minimum	Maximum

selecting appropriate locations for *liang pa'* cemeteries. In areas with high limestone cliffs, communal cemeteries were simply created in the closest and most structurally solid rocky faces. In areas with dispersed basaltic boulders, more locational options are offered, and distribution patterns may be affected by land ownership. Throughout Tana Toraja, therefore, rock faces have not been selected on the basis of cosmological concerns, but of availability, rock quality and the practical suitability of the site (Waterson 2009, 319).

But what exactly makes a rock face suitable for rock-cut tombs? Our study has highlighted two important aspects: first, there are several criteria, and secondly, these criteria have changed radically over time based on a comparison of the old and recent generations of *liang pa'*. This diachronic variability is summarised in Table 6.10 and Figure 6.12.

Before the Dutch arrived in Tana Toraja in 1906, rock-cut tombs tended to be located in discreet secluded landscape locations, and placed high up on cliffs in order to prevent looting or robbery. Such landscape locations also suited the nature of the

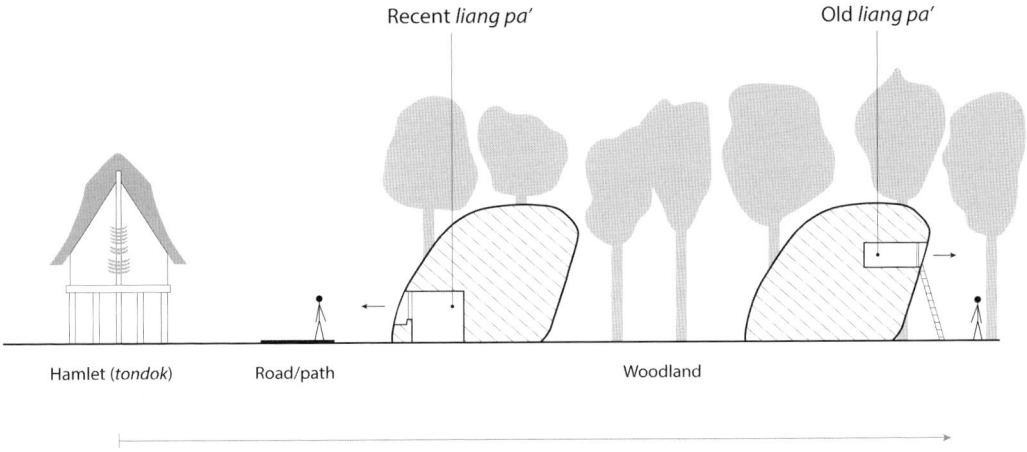

Figure 6.12: Diagram showing the differences in the location of old and recent liang pa' *rock-cut tombs in the landscape and in relation to settlements (drawing: G. Robin).*

social relationships between the living and the remains of the dead. Remote, hidden locations were selected as a way to provide respect and quietude to the dead bodies, which were to be visited and 'disturbed' only at specific occasions. In past traditional Toraja landscapes, 'villages of the dead' were clearly separated from those of the living, and the living were afraid of venturing near cemeteries which were considered as sacred, almost otherworldly places.

The decrease in warfare and violence after the arrival of the Dutch, the development of conversions to Christianity, and the boom in ceremonial expenditures from the 1970s onward, have modified the way locations have been selected for new *liang pa'* tombs (Fig. 6.12). Prioritising accessibility, social proximity with the dead, memory and wealth display, recent rock-cut tombs have been built close to roads and villages, at ground level. They are orientated to public view so that the dead can be remembered on a daily basis and new types of conspicuous (and expensive) rock decoration can be admired by all and therefore enhance the prestige of specific families (see Chapter 3). Improvements in stone-cutting techniques and in the design of tomb entrances have made recent *liang pa'* less vulnerable to rainwater and less dependent on naturally sheltered rock faces. This has provided stone cutters with more flexibility and has given more locational options to sponsoring families, contributing to the very dynamic character of *liang pa'* cutting practices today.

Notes

1. Note that the south is the back of the house, while its entrance is always on the north in order to face the sun. During the funeral ceremony, the corpse will exit the house from the north but its body orientation represents his/her symbolic departure toward the south.
2. Although, according to Nooy-Palm (1986, 183), stillborn are buried beneath the rice barn or to the east of the house.
3. Duli (2018, 43–44) also mentions an old cemetery with *erong* and *liang pa'* that is located *c.* 1500 m to the south of Bori': this might be boulder 10 in Parinding, which is associated with the villages of Parinding, not with Bori' (see map in Pl. 12).
4. According to one of our informants, if a *tongkonan* is moved to another location, soil from the original location must be moved along to the new location. This is because a *tongkonan* consists of not only the house structure, but also the land on which the *tongkonan* sits.
5. The avoidance of cemeteries in the past did not mean a lack of contact or relationship between the living and the dead. Waterson (2009, 207, 320–321) explains how Toraja people are characterised by continuous, intimate relationships with the dead and the ancestors through various rites and communications that are performed regularly outside cemeteries (*e.g.*, in houses) and which enable the maintenance of good relations with the ancestors as sources of blessing for their descendants.

Chapter 7

The social biography of *liang pa'* cemeteries

In Chapter 6, we demonstrated how cemeteries have developed through time, starting with old *liang pa'* and advancing progressively with the addition of more recent tombs, intensifying the use of old cliff faces and boulders, and expanding spatially over other cliffs and boulders in the vicinity. So far, we have concentrated on the physical environment of the cemeteries. In this chapter we explore the social environment of the cemeteries, looking at relationships between cemeteries and the people who used them, and how these relationships are established through space and time. The key principle here is the link between individual tombs and *tongkonan* kinship houses. Each tomb is paired with a specific *tongkonan*, and people associated with a *tongkonan* have burial rights in its *liang pa'*. This principle raises questions that have rarely been addressed in previous research. How many *liang pa'* can a *tongkonan* have? How far away can a tomb be from its *tongkonan*? Since Toraja individuals have membership in more than one *tongkonan* group, in which tomb do they generally prefer to be buried? This chapter explores the spatial relationships between *tongkonan* and cemeteries as places, using the district of Rembon as a case study. It then focuses on the cemetery of Sele in the district of Sesean Suloara' and investigates how genealogical developments of local *tongkonan* communities, and relationships between these communities, are mirrored in the biography of the cemetery.

Mapping kinship groupings: the geographical catchment of cemeteries

Liang pa', as durable funerary monuments, enable *tongkonan* groups to anchor their presence in the landscape, and to root their genealogy in safe, permanent rocky places. A *liang pa'* is always associated to a single *tongkonan* house of origin, the two being conceived as a 'pair' (*sipasang*). A *liang pa'*, therefore, cannot be associated to several *tongkonan*. A *tongkonan*, however, can have more than one tomb. One tomb will be created at the time of the foundation of the *tongkonan*, since 'no *tongkonan* is really complete without its *liang*' (Waterson 1995, 207). Other tombs can be added later, by the descendants of the original founder of the *tongkonan*. The commissioner of the *liang pa'* and their direct descendants have a right of burial in the *liang pa'*. As a consequence, an individual who is the direct descendant of several successive *liang pa'* commissioners have burial rights in several *liang pa'*.

The range of possible burial places for an individual is made even broader when one considers that Toraja people always have memberships in more than one *tongkonan*. The Toraja kinship system is bilateral (patrilineal and matrilineal ascendancies are equally important). In addition, marriage provides a link to the spouse's *tongkonan*, and marriages are both patrilocal or matrilocal (Waterson 2009, 174, 201–222). This means that each individual has an association to at least two *tongkonan* groupings (the father's and the mother's), and three if married (the spouse's *tongkonan*). Since the parents might also have several *tongkonan* affiliations themselves, an individual's inherited affiliations are multiple and often overlap across generations. As an example, Samuel Palangda', our guide in Rembon, said he could claim membership to 12 *tongkonan* groups and therefore, in theory, has burial rights in at least 12 *liang pa'*. According to Waterson (2009, 175), in practice, most individuals can only sustain (and therefore claim) ties to the *tongkonan* of their parents, grandparents and spouse's parents and grandparents (making a total of eight *tongkonan*, or fewer, as affiliations overlap).

How do people choose from these multiple burial options? To answer this question, one needs to consider the *tongkonan* to which individuals are more closely tied. This is mainly based on genealogical proximity (*e.g.*, one of the parent's *tongkonan*), and on the ability of the individuals to maintain ties with *tongkonan* groups, typically by regular participations at (and contributions to) ceremonies associated with the rebuilding of parts of, or whole, *tongkonan* origin houses, or at funerals of the *tongkonan* members. However, as *tongkonan* membership confers rights to the shared land and wealth of the group, people may be tempted to claim membership in genealogically more distant *tongkonan* groups with more resources, and leading members of these groups may challenge such claims (Waterson 1995). In practice, as far as we could observe in the field, where we were informed of the genealogical relationships of people buried inside individual tombs, it seems that the preference is to be buried in the *liang pa'* of one's own parents or siblings.

The place of burial is not determined by the individual's residence, nor predictable by any dominant *tongkonan* affiliation. It is not uncommon for two spouses and two siblings to be buried in different tombs (Waterson 1995). The sister of Samuel Palangda', for instance, was not buried in one of the two *liang pa'* that belong to her father's *tongkonan* (Mabarre'), which are located in the cemetery of Batu Lappa' (Pl. 15), nor in the modern *patane* built nearby by her brother. Instead, she wished to be buried in another cemetery located 5 km away, in the *lembang* of Ratte Talonge' (district of Saluputti), in a *liang pa'* that belongs to her mother's *tongkonan*. Therefore, tomb locations are not an index of individual mobilities (people do not create tombs where they have settled to live, unless they create a new *tongkonan*); they are rather an expression of kinship spatial immovability (people return to one of their *tongkonan*'s heartland to be buried). This means that the essence of kinship is deeply rooted geographically: the space in which the *tongkonan* and its tomb(s) are located constitute an important part of the kinship's identity. We may then deduce that a *liang pa'* should be relatively close to its *tongkonan*, *i.e.*, it should be created in the closest cemetery location available around the origin house. As a consequence, one can predict that a cemetery only has tombs of local *tongkonan* and not of distant ones. Does this principle apply to real situations observable in the field? If a region has several cemeteries, how do *tongkonan* decide the cemetery in which to create a tomb? What is the average distance between a *tongkonan* and its tombs? How wide is a cemetery catchment geographically?

To answer these questions, let us stay in the district of Rembon (our study area C). Three long-established cemeteries are located in the northern part of the district: Sanduni' (*lembang* of Ullin), Batu Lappa' (*lembang* of Buri') and Salu Liang (*lembang* of Kole Sawangan) (Pl. 16). The respective distances between these three cemeteries are 0.8 km (Sanduni to Batu Lappa'), 2.2 km (Batu Lappa' to Salu Liang), and 3 km (Salu Liang to Sanduni) as the crow flies. During our visit to the region in 2017, we collected the names of *tongkonan* origin houses which had tombs in the first two cemeteries (Sanduni' and Batu Lappa'). Information was provided by our guide Samuel Palangda', his father Abraham Sulu', Markus Tandikaloden (younger brother of Samuel's mother), Daud Moning Tulak (a friend of Samuel's family) and Paulus Kondosara, an 82-year-old ritual practitioner, comparable to a traditional *to minaa* ritual specialist (although he is Christian and is therefore not officially a *to minaa*). We were able to locate ten of the *tongkonan* associated with the cemetery of Sanduni' (Bamba, Bunna, Karassik, Layuk, Lameme, Madao, Oppon, Papa Batu, Rarung, Suka), and six of those associated with Batu Lappa' (Bamba, Bunna, Buttu, Durian, Mabarre', Papa Batu) (Pl. 16). Although this survey was not comprehensive (many more local *tongkonan* have a *liang pa'* in these two cemeteries), the sample data we collected is useful to show general geographical trends.

The first finding is that all the *tongkonan* houses are located within a 2.5 km radius from their respective cemetery, *i.e.*, an hour's walk maximum. It provides us with interesting examples of 'cemetery catchments'. Moreover, we have confirmation that all the *tongkonan* have used the geographically closest cemetery to create their *liang pa'*, as predicted. Another interesting finding is that two *tongkonan* origin-houses (Papa Batu and Bunna) each have a *liang pa'* in both Batu Lappa' and Sanduni' cemeteries, and not in only one cemetery, as was the case for the other houses in our survey. *Tongkonan* Papa Batu and Bunna are geographically equidistant from both cemeteries, so their 'in-between' geographical location may explain this particular feature. Papa Batu was founded six generations ago by a person called Buttu Batu and is the oldest *tongkonan* in the area. It has the peculiar distinction of having a roof made of stone slabs instead of bamboo, hence its name (meaning 'stone roof' in Toraja) (Fig. 7.1).

The stone slabs measure 50–60 cm long, 30–40 cm wide and 5–10 cm thick. This is a unique feature in Tana Toraja, which Paulus Kondosara (himself a direct descendant of Buttu Batu) explained as a wealth display from the founder of the *tongkonan*. The original *liang pa'* of Papa Batu was made by Buttu Batu, the founder of the *tongkonan*, in the cemetery of Sanduni'. Later, a descendant named Kaluden married into the *lembang* of Buri' and created another *liang pa'* in the cemetery of Batu Lappa' (which was closer to its residence but not further from *tongkonan* Papa Batu than is Sanduni': see map in Pl. 16). The place where *tongkonan* Papa Batu was built is called Tumakke. As a consequence, the *tongkonan*'s tombs in Sanduni' and Batu Lappa' are called *liang to Tumakke* ('the tomb of the people from Tumakke'), even though most people buried in them did not live in Tumakke.

Finally, this study in Rembon revealed an unexpected case of a single *liang pa'* being affiliated to two different *tongkonan* origin-houses. This oddity is due to particular circumstances. Several generations ago, a person called Rombe Salu had membership in *tongkonan* Rarung from his mother and in *tongkonan* Lameme from his father. During his life, Rombe Salu sponsored the entire rebuilding of the two *tongkonan* houses himself (while normally the costs are shared by the members of each *tongkonan*), as well as the creation of a new *liang pa'* in the cemetery of Sanduni' for himself and

Figure 7.1: Tongkonan *Papa Batu* ('stone roof') in Banga (Rembon) with its unique stone-tiled roof (photo: G. Robin).

his descendants. As a result, this tomb (called *liang Rombe Salu*) was associated with both *tongkonan*.

The social life of cemeteries

In Chapter 6, we have shown how the location of the tombs on the physical rock faces matter. In particular, we demonstrated that older *liang pa'* are more frequently located in higher locations (requiring a ladder), and in overhanging, sheltered rock faces, while more recent *liang pa'* are made closer to ground level and in sloping rock faces. Are there any further principles that have influenced the location of tombs within the micro-topography of the rock faces (on either cliffs or boulders)? Were specific areas on rock faces deemed 'better' or more appropriate, leading to some form of competition or negotiation between different *tongkonan* groups sharing the same rock cemetery?

According to Nooy-Palm (1979, 259), 'the higher a person's status, the higher the final resting place hewn out for him in the cliffside' (see also Koubi 1982, 196, 399, for a similar statement). Although such a principle may have prevailed in specific places, such as the high limestone cliff cemeteries of Lemo, Londa or Suaya, it is not prominent in most areas and generally there does not seem to be any particular 'hierarchy principle' in the layout of tombs in a cemetery. Paulus Kondosara confirmed that in Rembon, tomb locations in relation to each other on rock faces have no particular social meaning or importance.

How does a typical cliff cemetery develop over time? How do tombs gradually spread over the surface of the rock? These related issues have also not been addressed in the literature, although Kruyt (1924, 162, 167) postulated that the first tombs created in cemeteries would take the more accessible places in the rock face, while later tombs had to use the remaining, less accessible upper spaces. The cemetery of Sele in Suloara' (district of Sesean Suloara') provides an interesting case study. It consists of a single, large volcanic outcrop (40 m wide, 12 m high) in which a total of 35 *liang pa'* were created (Pl. 17). Sele is not the oldest cemetery in the area. 150 m to the south of Sele is the old cemetery of Pana', a vertical limestone cliff approximately 30 m high (Fig. 7.2). Pana' was first used as an *erong* cemetery, with wooden sarcophagi safely placed inside a horizontal crevice at the top of the cliff. In a subsequent phase, 32 *liang pa'* were gradually added on the cliff face underneath the crevice. All were created before Dutch colonisation, *i.e.*, between c. 1700 and 1900. The cemetery is no longer in use today. The last interment was made about 80 years ago and was placed in the crevice (*lo'ko'*-type burial), not in one of the *liang pa'*. It is not clear why the cemetery of Pana' ceased to be used, since there is still space for additional tombs on the cliff. It appears as though the large outcrop of Sele progressively 'took over' the role of local cemetery. This was certainly for practical reasons. In the past, when raids and warfare were common, Pana' was convenient as a hidden place with a very high cliff that enabled the creation of tombs that were safe from Bugis attacks or looting.

Figure 7.2: The old, disused cemetery of Sele in Suloara' (Sesean Suloara') (photo: G. Robin).

With the Dutch presence from 1906, these threats dissipated and people started to look for more accessible places for their tombs (Keers 1939, 207), which were also more visible to all (prestige display). The outcrop of Sele is located on the side of a path and aligns better with these new purposes.

The 35 *liang pa'* at Sele are all on the same rock face, which faces northeast (Pl. 17). The first *liang pa'* at Sele was created about 100 years ago, and the cemetery is still being used today with regular burials and *ma'nene'* rituals. Concrete steps on the western end of the outcrop were added in 2013. In June 2017, there were 26 tombs in use, four tombs that were fully hewn out but had not been consecrated and used yet (therefore left open with no door), and five future tombs not already cut but marked on the rock face as 'reserved spots'. Local elders Julius Kamma (aka Papa Kiki) and Ne' Sampe provided information that helped us trace the associated *tongkonan* and the construction periods for most of the tombs, as well as the individuals buried in some of them.

Tomb 31 in our plan (Pl. 17) is probably the oldest *liang pa'* in the cemetery. It was made more than 100 years ago according to our informants. Tombs 8, 15, 22, 24, 26, 31 and 35 were likely made during the same initial phase, prior to the 1950s, as they have several characteristic features of the old-style *liang pa'* (small entrance, no stone sill under the wooden door, old-style unpainted wood carvings, high position on the rock face). Tomb 2 was hewn out around 1950 but had not been used by the time of our visit. Tomb 10 was also created around 1950 but was used shortly afterwards. The last burial in this tomb took place in 2015. Other tombs were created later, in the 1960s (tomb 18), 1980s (tomb 6), 1990s (tomb 3) and 2000s (tombs 17, 29, 33). The latest tombs to be cut at Sele were tombs 4 (2011) and 1 (2015). Although incomplete, this chronological overview indicates that the first tombs at Sele used spaces across the entire width of the rock face, but not under a minimum height of 2 m above ground level. Subsequent tombs were cut in-between the first-generation tombs and above them. The last generation of tombs (those created from the 1990s onwards, plus the 'reserved spots') are located on the upper, lower and side margins of the rock face, including the spaces directly accessible from ground level.

The Sele outcrop is on land that belongs to *tongkonan* Rante Bulaan, whose origin-house lies 400 m to the northwest (Pl. 18). The cemetery, therefore, belongs to this *tongkonan*; however, it is shared with seven other local *tongkonan* groups whose origin-houses are located within a 500 m radius (except *tongkonan* Tiroan, which is 1.1 km away to the south). Because the cemetery of Sele is privately owned, not all local *tongkonan* are granted a right to create a *liang pa'* in it. Instead, only *tongkonan* groups that have close ties with the owners (*tongkonan* Rante Bulaan) have been allowed to do so. For instance, six *liang pa'* were made by members of *tongkonan* Toyasa, which has many members in the area and has also intermarried with several members of *tongkonan* Rante Bulaan. *Tongkonan* Ta'Pan Kila', which also has six tombs, was a *tongkonan layuk* ('great *tongkonan*'), i.e., the origin-house of a dominant kinship group that held political responsibilities in the area. Interestingly, tombs from the same *tongkonan* tend to be located close to each other within the rock face, creating slightly loose spatial clusters: for instance, the tombs of *tongkonan* Toyasa are mainly located on the eastern part of the outcrop, while *tongkonan* Ta'Pan Kila' occupies its centre, and *tongkonan* To Tallang has its three tombs on the western end of the outcrop. This spatial proximity among tombs of the same *tongkonan* might be interpreted as another expression of Toraja noble families' desire to not have their deceased members intermixed with those of other kinship groups.

Conclusion

The examples examined in this chapter show different social dynamics at play around and within *liang pa'* cemeteries. The first case study reveals how two neighbouring communal cemeteries are managed and shared by local *tongkonan* groups. The typical 'catchment area' of these cemeteries has a 2.5 km radius. Each kinship group uses the communal cemetery most closely located to their origin house. *Tongkonan* houses located equidistant between two cemeteries may have a tomb in both of them.

Our second case study, the cemetery of Sele, is a privately owned cemetery. It shows that only a limited number of local *tongkonan* are allowed to create a tomb in such cemeteries, based on their ties with the *tongkonan* owning the rocky place. As a result, the 'catchment area' in this case is smaller (0.5–1.0 km radius). Information on the creation date of the tombs at Sele provides insights into the spatial development of a *liang pa'* cemetery within a single rock face. Here, we observed a development starting from the centre of the rock face and finishing at its margins. Although the cemetery of Sele started to be used after the beginning of the Dutch colonial period (which ended raids and looting from neighbouring kingdoms), the first tombs maintained the traditional elevated location preference (beyond reach from ground level, requiring a ladder), which is typical of older generation *liang pa'* across Tana Toraja. This indicates that an elevated position was not only related to safety concerns, but likely to a need to maintain a certain distance between the dead and the living, or simply to build the house of the dead off the ground (as the houses of the living on its posts), and perhaps also to display social rank (building elevated tombs being more expensive as it requires bamboo scaffolding). The layout of the tombs in a cemetery cliff, therefore, is not totally random (*contra* Grubauer 1913, 200). The position of the tombs within the rock face and in relation to the other tombs *does* matter. With the addition of tombs, *tongkonan* groups invest rock faces with their social representations and concerns. The kinship-based clustering of tombs evidenced at Sele shows that not only the burial chambers but also the rock around the chambers become part of the inheritance and identity of *tongkonan* groups.

Chapter 8

Conclusion

In this book, we have aimed to provide a comprehensive overview of the *liang pa'* rock-cut tomb tradition of the Toraja people from Sulawesi, exploring their cultural context, use and signification. This was done using both existing ethnographic literature and our own fieldwork observations. Additionally, this book presents the main results of our fieldwork investigation carried out in June 2017 in Tana Toraja, geared towards addressing specific issues that have been overlooked in previous studies. In Chapter 1, we outlined seven key issues, which we can briefly revisit now.

The social value of rock-cut tombs

The Toraja rock-cut tomb tradition is rooted in the Austronesian cultural association between stone and the dead. In Tana Toraja, the dead of noble rank have long been deposited in dramatic rocky places (caves, rock-fissures, wooden sarcophagi hung on cliffs), conferring them with prestige. Rock-cut tombs represent a relatively recent development of this broader cultural framework. The practice of cutting tombs in rocky places likely did not begin until the 17th century AD. Rock-cut tombs are also unique to the Toraja culture of Sulawesi; no other rock-cut tomb traditions are known in Indonesia or the wider Austronesian world, where many megalithic tomb traditions are known instead.

In Tana Toraja, rock-cut tombs are part of a long-standing tradition of lavish burial and associated funerary practices. In this context, older aristocratic burial traditions, such as the *erong* sarcophagi, have fallen out of use, while *liang pa'* rock-cut tombs and *patane* house-tombs continue to be actively maintained traditions today. Nevertheless, rock-cut tombs have emerged as the preferred burial method over time and are more valued than *patane* by Toraja people because of the prestige associated with stone as a durable material.

The emergence of the rock-cut tomb tradition in the 17th century is often presented as a response to raids and looting from neighbouring groups. Arguably, rock-cut tombs also represented a way to increase prestige expression in Tana Toraja. Not only are rock-cut tombs found in prestigious rocky places, but they are also testimony to the extraordinary costs associated with hewing rectangular spaces in rock cliffs and outcrops. It is unclear why no megalithic-tomb tradition developed in Toraja culture, considering that Toraja people quarry and use standing stones (*simbuang batu*) in relation to both funerary and domestic rituals (Adams and Robin 2022) and that megalithic tomb traditions are prominent in other Indonesian islands (Steimer-Herbet 2018). However, unlike many other Indonesian islands, Sulawesi (and the Toraja highlands in particular), offers dramatic rock cliffs that have been associated with death since the origins of the Toraja culture. This combination of original landforms and cultural perceptions of the landscape in Tana Toraja provided the context for the emergence of rock-cut tombs. Eventually, rock-cut tombs, together with the elaborate funeral ceremonies of the nobility, have provided similar means of prestige and conspicuous display than megalithic tomb practices in other areas of the Austronesian world.

The architecture and decoration of rock-cut tombs

Rock-cut tombs are simple architectures, consisting of a basic rectangular space long enough for the deposition of extended bodies placed side by side and on top of each other. One outcome of our fieldwork has been the identification of two main styles of tombs corresponding to two successive generations in the development of the *liang pa'* tradition. The two styles display differences in both architecture and decoration. Older tombs (style 1, c. 1700–1960s) are smaller in size (typically 1 × 1 × 2 m) and are closed by a wooden frame-and-shutter system that was incised with traditional Toraja motifs, which do not appear to have been painted. Recent tombs (style 2, 1970s–today) are larger (typically 2 × 2 × 2 m) and some of them include an extra space in the form of an elevated recess large enough to contain the body of the commissioner of the tomb. Recent tombs also have a more elaborate rock-cut entrance which includes a rock threshold and side panels, onto which a large wooden shutter is fixed. In recent tombs, motifs are mainly found on the wooden door, where they are incised and painted; however, certain tombs also include reliefs carved into the rock around the entrance of the tomb. These carved motifs include buffalo heads, portraits of the founder(s) of the tomb and other traditional ornamental motifs associated with Toraja nobility. In both generations of tombs, decoration is limited to the external parts of the tombs, and is therefore primarily intended for the living as reminders of the noble rank of the kinship and people associated with the monuments. Both styles of tombs have a recessed entrance with a stone bench in front, which is used for the deposition of offerings to the dead.

In Toraja oral tradition, *liang pa'* are referred to houses of the dead (*tongkonan tangmerambu*, 'houses without smoke'). This conceptual association is not explicitly materialised in the architecture of the tombs. The architecture of the tombs is restricted to its main functional purpose of containing dead bodies and does not include formal, structural references to *tongkonan* architecture – the only exception we could observe were two tombs in Bori' that had fully carved *sembang* (house beam extensions). The rituals associated with the hewing phases of the tombs, however, directly reference house construction rituals. Moreover, the motifs used in the decoration of the wooden doors in both old and recent generations of tombs are directly taken from the repertoire of *tongkonan* wood carvings, and both are made by the same specialist artisans (not by the stone workers who cut the tombs). The range of motifs used in tomb decoration, however, is more selective than the wider repertoire of house decoration, as only four main motifs are used: buffalo head (*pa'tedong*), sun disk with rays (*pa'barre allo*), banyan tree leaves (*pa'barana' rapa'*) and basket lid motif (*pa'kapu' baka*). There are no decorative motifs used that are not derived from the house repertoire, except for the recent carved stone human portraits that can be regarded as modern versions of the traditional *tau-tau* effigies of the dead, which were displayed outside at the entrance of rock-cut tombs.

The construction of rock-cut tombs

Each rock-cut tomb is associated with a specific *tongkonan* kinship house located in the vicinity of the cemetery. A rock-cut tomb can be created by the founder of a new *tongkonan* or by its descendent. The creation of a new *liang pa'* is not dictated so much by practical consideration (the need for a new tomb to take over from existing ones that are full) as by social drivers: the prestige an individual can gain by sponsoring the creation of a new tomb. Tombs are always cut during the lifetime of the person who commissioned them and are named after them. Tombs may be complete long before the death of their sponsors and remain open until a first burial takes place, which occasions a consecration rite and the closure of the tomb. The first person buried inside a new tomb may not be the commissioner but another individual directly related to the commissioner (*e.g.*, spouse, sibling, child). One individual normally does not commission more than one tomb during their life.

Rock-cut tombs are made by specialised artisans, mainly based in villages located on the slope of Mount Sesean, in the northern part of Tana Toraja. These stone artisans also specialise in quarrying menhirs and carving stone pillars serving as bases for central front posts (*tulak somba*) of traditional *tongkonan* houses. Stone artisans are commissioned for their work. They were traditionally paid in buffaloes, while cash is now the most common form of payment. These artisans typically work in crews of three individuals, who rotate among the various tasks associated with cutting a *liang pa'* tomb. Metal picks of different lengths are used, with longer ones for rougher rock breaking and extraction, and shorter ones for more precise cutting work. The entire process, nowadays, takes 2–8 months.

The process of hewing a tomb comprises a series of successive technical stages, each initiated by a specific ritual. These rituals are performed by the stone artisans. They involve the sacrifice of animals and were traditionally intended for the spirits of nature (*deata*) to ensure that the intrusive process of rock hewing progresses successfully.

Studying the hewing process of tombs enabled us to observe the fate of waste rock, an aspect often overlooked in archaeological studies of rock-cut architecture. Rock-cut tombs also work as quarry sites. In Tana Toraja, blocks and chips extracted from rock-cut tombs are left outside the monuments by stone workers and are available for use by local communities during the time of the hewing process. A part of the waste rock remains in front of the tomb; another may be taken away and used, for instance, to create terrace facings. Interestingly, once a tomb is ritually consecrated, it is no longer permitted to take away waste rock extracted from it.

Burial practices

The ethnographic literature provides detailed descriptions of the complex body treatments that take place before and during funeral ceremonies of the Toraja nobility. Deceased individuals are gradually mummified and wrapped in multiple layers of cloth (*mebalun*) that take the shape of cocoon-like cylinders. The main intention of this process is to ensure that deceased individuals are preserved as articulated bodies and that the individual's remains do not become disarticulated and mixed with the bones of other individuals in the collective rock-cut tombs. Once such treatment is complete and funeral ceremonies have taken place, bodies can be placed inside the tombs. Bodies are typically placed head first, with the feet closer to the entrance of the tomb. The full biography of long-used individual tombs (from first interment to disuse) is difficult to establish in the field; therefore, it was not possible to determine the maximum number of bodies rock-cut tombs can contain. However, notional figures could be established and these varied across the two generations of tombs we identified.

Older tombs (style 1) were designed to typically contain up to 10–12 wrapped bodies, with multiple rows of two bodies side by side. The *ma'nene'* ritual, practised in different parts of Tana Toraja, involves the extraction and rewrapping of ancient interments, which means that wrapped bodies are curated over several generations. However, after a certain time, which we could not establish, bodies are no longer curated and are left to decay. As a result, an old disused tomb often presents bodies in different states of preservation. The upper (latest) depositions are individual bodies wrapped in cloth, while the lowest (oldest) depositions consist of comingled bones of multiple individuals. The comingled bones get naturally fragmented over time, thus creating space for further burials even in tombs that were once considered full. This means that, theoretically, older tombs could be used indefinitely, although in practice most have been disused with the development of most recent rock-cut tombs.

Recent tombs (style 2) have the same length as older ones (2 m) but are wider and taller. This enables a larger number of body depositions, in theory up to 20. However, according to our observations, these tombs are rarely used to full capacity, due to multiple factors (*e.g.*, the high number of tombs available to families and the concomitant use of *patane* tombs). We also noted that the traditional mummification process involving cloth wrapping is being gradually replaced by the injection of embalming preservatives (formaldehyde) into the bodies, which are subsequently placed into individual coffins before being deposited inside rock-cut tombs.

All the deceased placed inside a rock-cut tomb belong to the same *tongkonan* kinship group. Within the tombs, the relative position of bodies is not determined by rank, age or gender. In fact, there is no differentiation in body placement, except in rare cases, such as recent tombs that include a separate recess space for the sponsor of the tomb, or larger tombs with two separate entrances (which are divided into two spaces, one for each family).

Interactions with the bodies of the dead continue after burial. These can be direct interactions, such as the *ma'nene'* ritual mentioned above, or indirect, with the regular deposition of offerings on the entrance recess of the tombs. These offerings traditionally consisted of betel nuts, but nowadays they take the form of water bottles, cigarettes, banknotes and others. No offerings are placed inside the tombs directly with the bodies (but valuable body ornaments may be wrapped together with the body during the funeral ceremony and thus be present inside the tomb – hence the temptation, severely punished, to loot them in the past).

The landscape setting of rock-cut tombs

Cardinal directions are key structural principles of the ritual life of Toraja people. Practices associated with the traditional religion (*aluk to dolo*) are divided into the Rites of the East (concerned with life and reproduction) and the Rites of the West (concerned with death and the ancestors). In Toraja cosmography, the world of the dead (Puya) is located somewhere far to the south. Many rituals associated with funerals are directly and explicitly associated with western and southern locations of, and orientations from, the domestic space (house, village). One could expect that such cosmographic principles could have influenced the location of cemeteries in relation to villages. Indeed, several ethnographic works have asserted that cemeteries were founded to the west or south of villages. Our field observations, however, show that this was not the case. The choices for selecting suitable locations for the creation of cemeteries were mainly guided by practical concerns of security, with a clear preference for high cliffs located in forested, secluded areas of the landscape. The cardinal location of a cemetery in relation to its founding village did not matter; in fact, a cemetery is often used by several village communities distributed in various locations in the surrounding vicinity. In areas on the slopes of Mount Sesean, where the practice of tomb hewing is the most active today, tombs are mainly created in small volcanic boulders that are scattered across the landscape: in this region, *tongkonan* members create tombs in boulders located on the *tongkonan*'s own private land, typically rice paddies.

Patterns, however, can be observed in the distances separating settled areas and cemeteries. Our study of neighbouring communal cemeteries in the district of Rembon shows that each cemetery has an average catchment radius of 2.5 km, representing an hour's walk maximum from a *tongkonan* to its associated *liang pa'*. On the other hand, a tomb cannot be too close to settled areas; for instance, it cannot be located within a *tondok* (hamlet) or in the same land parcel as its corresponding *tongkonan*.

Geology and landforms have an impact on rock-cut tomb distributions and locations. In areas dominated by limestone geology, tombs are concentrated on a few large cliffs which can host dozens of tombs. In volcanic regions of Tana Toraja, tombs are found in smaller clusters within multiple boulders scattered across the landscape, creating a more widespread distribution overall. For the same reasons, the dynamics of the spatial development of cemeteries over time differ in the limestone-dominated and volcanic regions of Tana Toraja. In the former, cemeteries intensify over time, with the addition of new tombs on the same large rock face. In the latter, cemeteries start with a small group of tombs made on a large rock boulder; subsequent tombs are created in other boulders located immediately around the initial core boulder, and then again to an outer circle of boulders, thereby creating a concentric development over time (*e.g.*, Parinding and Bori' in Sesean).

Surprisingly, rock type does not influence the cost of hewing a tomb, at least today, as our study shows: creating a tomb in a (hard) volcanic boulder or a (soft) limestone cliff costs the same. However, the morphology of boulders and rock faces is taken into account when choosing a location for a new tomb. This was particularly significant in the past. Tombs of the older generation (style 1) are almost always located on high cliffs (offering locations only reachable with a ladder) presenting sub-vertical inclinations (offering shelter to the entrance of the tombs).

Cemeteries and kinship systems

Rock-cut tombs are exclusively associated with nobility and kinship in Toraja culture. Each tomb is paired with a *tongkonan* kinship house, and individuals with membership in a *tongkonan* kindred group can claim a right of burial in its *liang pa'*. The founder of a new *tongkonan* is also the founder of its associated *liang pa'*; moreover, descendants of the founder commonly create additional rock-cut tombs over generations, thus increasing the 'pool' of tombs available to members of the *tongkonan*.

The rock-cut tombs of a *tongkonan* are normally found in the same cemetery, typically the closest available from the kinship house. Medium-large cemeteries are often shared by several *tongkonan*. Our study of the cemetery of Sele in the district of Sesean Suloara' shows that tombs from the same *tongkonan* are often grouped together within the rock face. The position of tombs on the rock face, however, does not correspond to differences in social status between local *tongkonan* groups.

Impact of colonial and post-colonial changes on rock-cut tomb practices

The Toraja territory was under Dutch colonisation between 1909 and 1942. Colonisation has had two main long-term effects: a decrease in violence (both intrusive and intracommunity) and the spread of the Christian religion (gradually replacing traditional Toraja religion). In addition, from the 1960s, Indonesia experienced significant economic development, manifested in Tana Toraja by the adoption of cash crops, all resulting in an increase in ceremonial expenditures associated with funeral ceremonies and tomb building. Altogether, these political, religious and economic changes have impacted rock-cut tomb practices in Tana Toraja, as shown by our comparative analyses of older (style 1) and recent (style 2) generations of *liang pa'*.

A first noticeable consequence has been the sharp increase in the creation of rock-cut tombs in recent decades, especially in the northern part of Tana Toraja, where numerous volcanic boulders offer a large choice of rock faces for new tombs. In that region (our study area A), recent tombs represented 62% of the whole. More significantly, the landscape setting and the visual impact of rock-cut tombs have changed dramatically between the two generations. The locational preferences of older rock-cut tombs were mainly dictated by security concerns. As a result, tombs were placed on high cliffs located in isolated, secluded parts of the landscape, often forested areas away from road networks. Where boulders were on less secluded locations, tombs were made on the faces looking away from paths and settled areas. The entrances of the tombs were small, and their engraved wooden doors unpainted, thus having a limited visual impact. Cemeteries were isolated, avoided places associated with supernatural powers, which people would only visit on special occasions and only during a short period of the year (the funeral season, August–September). By contrast, recent tombs are more concerned with social display, memory and accessibility. They are made on rock faces exposed to public view, in particular on the side of roads and open landscapes. They have large, colourful wooden doors. Some tombs display large, stone-carved reliefs painted with bright colours, which represent additional construction costs. Finally, recent rock-cut tombs are often located on the lowest part of rock faces, at ground level, enabling more regular interactions (for instance during Christian festivals throughout the year), which in the past would not have been permitted.

Toraja rock-cut tombs offer a fascinating case study of a long-standing monument tradition adapting to major concomitant social changes. While the primary functions of *liang pa'* have remained unchanged (collective burial monuments associated with noble kinship groups), their emphasis has shifted from protecting and sacralising the dead to displaying and commemorating their status (and that of their descendants).

***

Our research has left several questions unanswered and has created new ones. There is certainly scope for more fieldwork investigation around *liang pa'* and other burial monuments in Tana Toraja. Our own short-term research has explored small sample areas, with a particular attention to the northern districts of the Mount Sesean area. Other areas not included in the present study (for instance districts south of Makale) may show different patterns than those presented here. Further, more marginal areas, located far from the main towns of Rantepao and Makake, have likely been less impacted by modern changes (including tourism) and would be worth investigating too. In other words, it would be of great benefit to expand research geographically to develop a broader and more representative dataset to address the general questions examined in this book, such as the distribution, landscape setting, architectural and decoration diversity of rock-cut tombs. This could be coupled with a dating programme to tackle the uncertainty regarding the date of emergence of the *liang pa'* tradition, and its chronological relation to other traditions such as *erong* sarcophagi and *patane* house-tombs.

Intensifying investigations on targeted case study locations, comprising a single cemetery site and its associated *tongkonan* kinship groups, would also be promising. Such an approach could establish comprehensive and detailed genealogies (founders of *tongkonan* and their descendants) and to correlate them with the biographies of individual tombs and the entire cemetery. This would undoubtedly illuminate questions we have only touched upon superficially here, especially about the traditional proscriptions and individual choices surrounding the creation of new tombs and the distribution of individuals in multiple kinship tombs. We see this book as a starting point. We hope it will contribute to raise awareness of this unique Indonesian tradition and heritage and encourage interest in its further study.

Finally, the research presented in this book has the potential to inform studies on ancient rock-cut architectures (tombs, temples, sanctuaries, churches, *etc.*), for which literary sources are often not available. Several of the issues pertaining to rock-cut monuments discussed here have often been overlooked in archaeological studies. For instance, the position of the monuments in the landscape (including the microtopography of the rock face), as well as their visibility and accessibility. These are important considerations for interpreting the social uses and functions of rock-cut monuments and, more broadly, the worldviews of the communities who created them. Rock-cut monuments are, first of all, the expressions of cultural perceptions of, and interactions with, the physical landscape. This may have multiple implications. For instance, ritual depositions found immediately outside prehistoric rock-cut tombs may not have been necessarily intended for the dead but, instead, for non-humans associated with the mineral world. Such offerings can pre-date the first use of the tombs and be associated with construction rituals.

More attention should be devoted archaeologically to waste rock material associated with the hewing process of these architectures. It is notoriously difficult to date directly the time of construction associated with rock-cut architectures (unlike built architectures) using traditional archaeological methods, such as radiocarbon dating. Deposits found inside rock-cut monuments can provide dates pertaining

8. Conclusion

to their (latest) use, but not for their construction. A stratigraphic study of waste rock deposits potentially still preserved outside the entrance of ancient rock-cut monuments offers scope for precisely dating the hewing phase(s) of the monuments.

Lastly, the ethnoarchaeology study presented in this book reminds us about the caution needed when interpreting the origin of comingled human remains found in past collective rock-cut tombs (*e.g.*, Shanks and Tilley 1982): comingled human remains, as those we could observe in long disused *liang pa'*, might not be intentional outcomes of burial practices, but rather unintended effects of long term taphonomy.

Our study of *liang pa'* cemeteries does not intend to bring ready-made answers applicable to rock-cut contexts from other periods and regions of the world. Instead, we hope it can help expand the range of questions and considerations that archaeologists, historians and anthropologists can use when approaching human practices of creating ritual architectures within solid rock.

Glossary of Toraja terms

Alang: rice barn.
Allo: sun, day, daylight.
Aluk to dolo: 'way of the ancestors', expression referring to the traditional Toraja religion.
Baka: large basket made from woven bamboo.
Bala'kayan (or *bala'kaan*): meat distribution platform used at high-ranked funeral ceremonies.
Ban manuk: 'to give chicken', rite involving bringing food for the dead at their rock-cut tombs.
Barana': banyan tree (*Ficus religiosa*).
Barre allo: sun disc with radiating rays (decorative motif).
Bate lepong: see *tau-tau lampa*.
Bombo: soul, spirit of the deceased.
Bombo dikita: 'the soul of the dead which can be seen', expression referring to the *tau-tau* effigies of the deceased.
Deata: deities.
Dibalik to mate: 'the deceased has been turned', ritual taking place at the beginning of funeral ceremonies, consisting of changing the orientation of the body of a deceased individual from east–west to north–south, and officially marking its death.
Dipopenlilli': 'regarded as foundations', refers to *pelilli* ('underlay'), slaves who were buried in the tomb of their master.
Dirambu: 'smoked', corpse treatment applied during funeral ceremonies.
Dirapa'i: highest rank of funeral ceremony.
Ditundan: awakening of the *tau-tau*, a ritual in high-ranked funeral ceremonies.
Di tutu'i: 'to close (with a door)', ritual associated with the placement of a wooden door on a rock-cut tomb.
Duba duba: long piece of white fabric used at high-ranked funeral ceremonies.
Dulang: wooden dish on a pedestal, used to serve food during ceremonies.
Erong: wooden sarcophagus hanging on cliffs.
Induk: palm tree (*Arenga pinnata*).
Issong: mortar made from a hollowed-out log, used for pounding rice.
Kabongo': life-size, wood-sculpted buffalo head fitted out with real horns, used in the decoration of *tongkonan* houses.
Kaunan: slave.
Kayu: tree; wood.
Kamiri: candlenut (*Aleurites triloba* or *A. moluccanus*).
Lakkean (or *lakkian*): corpse platform used at high-ranked funeral ceremonies.
Lamba lamba: long piece of red fabric used at high-ranked funeral ceremonies.
Layuk: great.
Lembang: district.
Liang: tomb.
Liang batu: stone tomb (another term for rock-cut tomb).
Liang pa': rock-cut tomb.
Liang piah: infant burial cut into a living tree.
Lo'ko': hole in the rock, cave.

Glossary of Toraja terms

Ma'batang: 'to make a tree trunk', expression to describe the process of tightly wrapping a corpse within a cylinder of cloths (*mebalun*); also the name of a ritual in funeral ceremonies.
Ma'bua: consecration ceremony of a *tongkonan* (also called *merok*).
Ma'bubung: ritual associated with the completion of the roof of a house or the ceiling of a rock-cut tomb.
Ma'nene': ritual involving the extraction of corpses from tombs in order to rewrap them with cloths, or the retrieval of *tau-tau* effigies from cemetery cliffs to change their clothes.
Ma'pakande tau-tau: 'to feed the effigy', a ritual in high-ranked funeral ceremonies.
Ma'pakande to mate: 'to give food to the dead', ritual involving bringing food at a *liang pa'* for the dead.
Ma'parampo: 'to arrive where one is intending to arrive', a ritual associated with the process of tomb hewing.
Ma'peliang: 'to put into the *liang*', entombment ritual.
Ma'rambu langi': ritual to appease the ancestors following a theft in a tomb (see *meloko*).
Ma'rapa' sarigan: 'to make touch' or 'to assemble the *saringan*', a ritual in high-ranked funeral ceremonies (also called *mesarigan*).
Ma'rapa' tau-tau: 'to assemble the effigy', a ritual in high-ranked funeral ceremonies.
Ma'rinding: 'to make the wall', a ritual associated with the process of tomb hewing.
Ma'sali (or *massali*): 'to lay out the floor', a ritual associated with the process of tomb hewing.
Ma'siku: a ritual associated with the process of tomb hewing.
Ma'suri: 'to reserve', a ritual associated with marking the location of a future tomb on a rock face.
Ma'tau-tau: 'to make the *tau-tau*', a ritual in high-ranked funeral ceremonies.
Manglassak tau-tau: ritual involving the addition of genitals to a *tau-tau* effigy in high-ranked funeral ceremonies.
Manglelleng sarigan: 'to cut [the wood for] the *saringan*', a ritual in high-ranked funeral ceremonies.
Manglelleng tau-tau: 'cutting [the wood for] the *tau-tau*', a ritual in high-ranked funeral ceremonies.
Mang lilli' liang: ritual consisting of placing the clothes of a slave person on the floor of the *liang pa'* of their master, so that the body of the latter can be placed on the slave's clothing.
Mangrambu bulisak: 'to blacken the wood chips with smoke', ritual associated with the preparation of the *saringan* in high-ranked funeral ceremonies.
Mangrara: 'to anoint with blood', consecration ritual.
Massabu: 'to consecrate, to inaugurate'.
Memangan: 'it is astonishing, wonder-inspiring'.
Meaa: 'to bury', ritual of entombment in *liang pa'*.
Mebalun: corpse wrapping ritual.
Meloko: transgression of stealing from a *liang pa'*.
Merok: consecration ceremony of a *tongkonan* (also called *ma'bua*).
Mesarigan: see *ma'rapa' sarigan*.
Naga: water-serpent decorative motif, symbol of the Underworld.
Nangka: jackfruit tree (*Artocarpus heterophyllus*).
Narang: horse.
Ne': shortening of *nene'*.
Nene': grandparent, ancestor.
Pa': chisel; cut, carved, chiselled out.
Pa'barana' (or *pa'barana'rapa'*): carved motif representing leaves of the banyan tree (*Ficus religiosa*).
Pa'barre allo: carved 'sunburst' or sun disk with rays (decorative motif).
Pa'bulu londong: carved motif representing cock feathers.
Pa'erong: geometric wood carvings made on *erong* sarcophagi.
Pa'kapu' baka: carved motif representing the tight bindings that secure the lid of the *baka* baskets.
Pa'sussu': woodcarving on *erong* and old *alang* consisting of incised vertical parallel grooves.
Pa'tedong: buffalo head motif.

Pande tau-tau (or *pande kay*): see *to ma'tau-tau*.
Pangala': woodland.
Parundun bombo: 'to bring food to the *bombo*', ritual involving bringing food at a *liang pa'* for the dead.
Patane: house-like tomb.
Pateng: cultivated land (rice).
Pelilli': slaves buried in the tomb of their masters to serve them in the afterlife.
Pemali: prohibition, taboo.
Pondan: pineapple plant; yarn made from the fibres of pineapple leaves.
Puang: lord.
Rante: plain, flat ground; ceremonial plaza.
Rapasan: temporary coffin used in high-ranked funeral ceremonies.
Sali: floor; middle room in *tongkonan* houses; platform under *alang* rice barns.
Salu: river.
Saringan (or *sarigan*): palanquin used in high-rank funeral ceremonies to transport the body of the deceased.
Sarong: large conical hat made from bamboo bark or rattan to protect from the sun and rain.
Sembang: upwardly pointing ends of the longitudinal beams in *tongkonan* houses and *alang* rice barns.
Sendana: sandalwood tree (*Santalum album* L.).
Simbuang batu: standing stone erected as part of high-ranked funeral ceremonies.
Sipasang: pair (formed by a *tongkonan* house and its *liang pa'* rock-cut tomb).
Sumbung: back room in *tongkonan* houses.
Tana' bassi: 'iron stake' (lower rank of Toraja nobility).
Tana' bulaan: 'golden stake' (upper rank of Toraja nobility).
Tau-tau: wood-sculpted effigy of the deceased.
Tau-tau lampa: temporary effigy of the deceased made with bamboo wrapped in cloth; also called *bate lepong*.
Tedong: buffalo.
To buda: 'the many' (commoners).
To dolo: the ancestors.
To ma'tau-tau: 'the one who makes *tau-tau*', maker of effigies of the dead; also called *pande tau-tau* or *pande kay*.
To makaka: freemen, lower rank of Toraja nobility.
To mangraruk (or *to umpabendan*): 'the one who erected', founder of a *tongkonan*.
To mangria sampin: female relatives of the deceased who supervise tasks at the funeral ceremony.
To mebalun: 'the one who wraps', slave responsible for wrapping corpses at funeral ceremonies.
To meloko: person who committed a *meloko* transgression (see *meloko*).
To minaa: 'knowledgeable person', ritual priest.
To pa'pa'na: 'the one who pierced', founder of a *liang pa'* rock-cut tomb.
Tondok: hamlet, village.
Tongkonan: origin house of a kinship group.
Tongkonan tangmerambu: 'house without smoke' (tomb).
Tulak somba: front post in *tongkonan* houses.
Umbaa kande (or *lao umbaa kande*): 'to bring food', ritual involving bringing food at a *liang pa'* for the dead.
Uru: tree (*Magnolia vrieseana* or *Michelia celebica*).

Appendix: list of cemeteries surveyed in June 2017

A. List of liang pa' cemeteries surveyed in June 2017

Study area	Rock ID (cemetery)	District (Kecamatan)	Domain (Lembang/ Kelurahan)	GPS location Latitude	GPS location Longitude	Rock type	No. liang pa'
		Regency (Kabupaten): Toraja Utara					
A	Batutumonga 1	Sesean Suloara'	Lempo	-2,910322	119,883376	Boulder	1
A	Batutumonga 2	Sesean Suloara'	Lempo	-2,910477	119,883399	Boulder	3
A	Batutumonga 3	Sesean Suloara'	Lempo	-2,914334	119,881505	Boulder	1
A	Batutumonga 4	Sesean Suloara'	Lempo	-2,914338	119,881434	Boulder	1
A	Batutumonga 5	Sesean Suloara'	Lempo	-2,914327	119,881257	Boulder	1
A	Batutumonga 6	Sesean Suloara'	Lempo	-2,914281	119,881233	Boulder	1
A	Batutumonga 7	Sesean Suloara'	Sesean Matallo	-2,914621	119,875067	Boulder	1
A	Batutumonga 8	Sesean Suloara'	Sesean Matallo	-2,914717	119,875	Outcrop	10
A	Batutumonga 9	Sesean Suloara'	Sesean Matallo	-2,914277	119,872495	Boulder	1
A	Batutumonga 10	Sesean Suloara'	Sesean Matallo	-2,914242	119,872471	Boulder	1
A	Batutumonga 11	Sesean Suloara'	Sesean Matallo	-2,914135	119,872329	Boulder	1
A	Batutumonga 12	Sesean Suloara'	Sesean Matallo	-2,914099	119,872251	Boulder	1
A	Batutumonga 13	Sesean Suloara'	Sesean Matallo	-2,914056	119,872248	Boulder	1
A	Batutumonga 14	Sesean Suloara'	Sesean Matallo	-2,914081	119,872222	Boulder	1
A	Batutumonga 15	Sesean Suloara'	Sesean Matallo	-2,914035	119,872192	Boulder	1
A	Batutumonga 16	Sesean Suloara'	Sesean Matallo	-2,912607	119,871138	Boulder	2
A	Batutumonga 17	Sesean Suloara'	Sesean Matallo	-2,911849	119,870317	Boulder	9
A	Batutumonga 18	Sesean Suloara'	Sesean Matallo	-2,911667	119,870124	Boulder	2
A	Batutumonga 19	Sesean Suloara'	Sesean Matallo	-2,91111	119,87003	Boulder	1

Study area	Rock ID (cemetery)	District (Kecamatan)	Domain (Lembang/ Kelurahan)	GPS location		Rock type	No. liang pa'
				Latitude	Longitude		
A	Batutumonga 20	Sesean Suloara'	Sesean Matallo	-2.910603	119.870599	Boulder	1
A	Batutumonga 21	Sesean Suloara'	Sesean Matallo	-2.910531	119.870586	Boulder	3
A	Batutumonga 22	Sesean Suloara'	Sesean Matallo	-2.910397	119.870398	Boulder	3
A	Batutumonga 23	Sesean Suloara'	Sesean Matallo	-2.915583	119.871814	Boulder	1
A	Bori'' 1	Sesean	Bori''	-2.920012	119.919685	Boulder	29
A	Bori'' 2	Sesean	Bori''	-2.919983	119.919594	Boulder	1
A	Bori'' 3	Sesean	Bori''	-2.920245	119.91969	Boulder	1
A	Bori'' 4	Sesean	Bori''	-2.919891	119.91934	Boulder	8
A	Bori'' 5	Sesean	Bori''	-2.919781	119.919163	Boulder	1
A	Bori'' 6	Sesean	Bori''	-2.919695	119.919125	Boulder	1
A	Bori'' 7	Sesean	Bori''	-2.919593	119.919018	Boulder	3
A	Bori'' 8	Sesean	Bori''	-2.919539	119.918964	Boulder	5
A	Bori'' 9	Sesean	Bori''	-2.919448	119.918991	Boulder	1
A	Bori'' 10	Sesean	Bori''	-2.919378	119.918948	Boulder	3
A	Bori'' 11	Sesean	Bori''	-2.919367	119.919034	Boulder	1
A	Bori'' 12	Sesean	Bori''	-2.919737	119.919533	Boulder	1
A	Bori'' 13	Sesean	Bori''	-2.919566	119.919694	Boulder	3
A	Bori'' 14	Sesean	Bori''	-2.920423	119.920685	Boulder	2
A	Bori'' 15	Sesean	Bori''	-2.920313	119.920672	Boulder	1
A	Bori'' 16	Sesean	Bori''	-2.920342	119.920755	Boulder	1
A	Bori'' 17	Sesean	Bori''	-2.92042	119.920806	Boulder	3
A	Bori'' 18	Sesean	Bori''	-2.920417	119.920886	Boulder	1

Study area	Rock ID (cemetery)	District (Kecamatan)	Domain (Lembang/Kelurahan)	GPS location		Rock type	No. liang pa'
				Latitude	Longitude		
A	Bori' 19	Sesean	Bori'	-2.920109	119.92069	Boulder	2
A	Bori' 20	Sesean	Bori'	-2.92008	119.920593	Boulder	1
A	Bori' 21	Sesean	Bori'	-2.92012	119.920888	Boulder	27
A	Bori' 22	Sesean	Bori'	-2.920029	119.920864	Boulder	2
A	Bori' 23	Sesean	Bori'	-2.919989	119.920896	Boulder	1
A	Bori' 24	Sesean	Bori'	-2.919885	119.921038	Boulder	3
A	Bori' 25	Sesean	Bori'	-2.919807	119.920936	Boulder	1
A	Bori' 26	Sesean	Bori'	-2.920061	119.92132	Boulder	3
A	Bori' 27	Sesean	Bori'	-2.919319	119.921258	Boulder	1
A	Bori' 28	Sesean	Bori'	-2.919327	119.920877	Boulder	2
A	Buntu Lobo 1	Sesean	Buntu Lobo	-2.932196	119.896399	Boulder	1
A	Buntu Lobo 2	Sesean	Buntu Lobo	-2.932137	119.896002	Boulder	3
A	Buntu Lobo 3	Sesean	Buntu Lobo	-2.931968	119.896608	Boulder	1
A	Buntu Lobo 4	Sesean	Buntu Lobo	-2.931894	119.896601	Boulder	2
A	Buntu Lobo 5	Sesean	Buntu Lobo	-2.931887	119.896683	Boulder	1
A	Buntu Lobo 6	Sesean	Buntu Lobo	-2.928318	119.896384	Boulder	1
A	Buntu Lobo 7	Sesean	Buntu Lobo	-2.928116	119.896344	Boulder	4
A	Buntu Lobo 8	Sesean	Buntu Lobo	-2.928097	119.896762	Boulder	1
A	Buntu Lobo 9	Sesean	Buntu Lobo	-2.928113	119.897041	Boulder	6
A	Buntu Lobo 10	Sesean	Buntu Lobo	-2.928086	119.897121	Boulder	2
A	Buntu Lobo 11	Sesean	Buntu Lobo	-2.927475	119.896955	Boulder	2
A	Buntu Lobo 12	Sesean	Buntu Lobo	-2.92748	119.897108	Boulder	1

(Continued)

Study area	Rock ID (cemetery)	District (Kecamatan)	Domain (Lembang/ Kelurahan)	GPS location		Rock type	No. liang pa'
				Latitude	Longitude		
A	Buntu Lobo 13	Sesean	Buntu Lobo	-2.927416	119.897025	Boulder	1
A	Buntu Lobo 14	Sesean	Buntu Lobo	-2.926274	119.895922	Boulder	1
A	Buntu Lobo 15	Sesean	Buntu Lobo	-2.926365	119.896622	Boulder	1
A	Buntu Lobo 16	Sesean	Buntu Lobo	-2.925902	119.896316	Boulder	1
A	Buntu Lobo 17	Sesean	Buntu Lobo	-2.920424	119.895222	Boulder	2
A	Buntu Lobo 18	Sesean	Buntu Lobo	-2.920617	119.89508	Boulder	3
A	Buntu Lobo 19	Sesean	Buntu Lobo	-2.920663	119.895048	Boulder	1
A	Buntu Lobo 20	Sesean	Buntu Lobo	-2.920374	119.895064	Boulder	4
A	Buntu Lobo 21	Sesean	Buntu Lobo	-2.920299	119.894941	Boulder	1
A	Buntu Lobo 22	Sesean	Buntu Lobo	-2.922279	119.900236	Boulder	1
A	Buntu Lobo 23	Sesean	Buntu Lobo	-2.922374	119.900295	Boulder	1
A	Buntu Lobo 24	Sesean	Buntu Lobo	-2.920724	119.901991	Boulder	1
A	Buntu Lobo 25	Sesean	Buntu Lobo	-2.919661	119.901919	Boulder	2
A	Buntu Lobo 26	Sesean	Buntu Lobo	-2.918332	119.902002	Boulder	1
A	Buntu Lobo 27	Sesean	Buntu Lobo	-2.918311	119.90212	Boulder	2
A	Buntu Lobo 28	Sesean	Buntu Lobo	-2.917419	119.902181	Boulder	1
A	Buntu Lobo 29	Sesean	Buntu Lobo	-2.917317	119.902106	Boulder	1
A	Buntu Lobo 30	Sesean Suloara'	Lempo	-2.917025	119.902688	Boulder	2
A	Buntu Lobo 31	Sesean Suloara'	Lempo	-2.916899	119.902559	Boulder	3
A	Buntu Lobo 32	Sesean Suloara'	Lempo	-2.916655	119.902943	Boulder	2
A	Buntu Lobo 33	Sesean Suloara'	Lempo	-2.916596	119.902881	Boulder	1
A	Deri 1	Sesean	Bori' Ranteletok	-2.91015	119.914802	Boulder	1

Study area	Rock ID (cemetery)	District (Kecamatan)	Domain (Lembang/Kelurahan)	GPS location		Rock type	No. liang pa'
				Latitude	Longitude		
A	Deri 2	Sesean	Bori' Ranteletok	-2.910193	119.914692	Boulder	7
A	Deri 3	Sesean	Bori' Ranteletok	-2.910249	119.91473	Boulder	1
A	Deri 4	Sesean	Bori'	-2.910308	119.914617	Boulder	5
A	Deri 5	Sesean	Bori'	-2.910346	119.914539	Boulder	1
A	Deri 6	Sesean	Bori'	-2.910429	119.914748	Boulder	1
A	Deri 7	Sesean	Bori'	-2.910528	119.914638	Boulder	1
A	Deri 8	Sesean	Bori' Ranteletok	-2.91053	119.915202	Boulder	1
A	Deri 9	Sesean	Bori'	-2.910272	119.914162	Boulder	2
A	Deri 10	Sesean	Deri	-2.912207	119.908057	Boulder	3
A	Deri 11	Sesean	Deri	-2.912449	119.907965	Boulder	2
A	Deri 12	Sesean	Deri	-2.912712	119.907017	Boulder	2
A	Deri 13	Sesean	Deri	-2.912862	119.906904	Boulder	5
A	Deri 14	Sesean	Deri	-2.91295	119.906877	Boulder	2
A	Lempo 1	Sesean Suloara'	Lempo	-2.908003	119.902791	Boulder	1
A	Lempo 2	Sesean Suloara'	Lempo	-2.910726	119.903401	Boulder	2
A	Lempo 3	Sesean Suloara'	Lempo	-2.907995	119.901769	Boulder	1
A	Lempo 4	Sesean Suloara'	Lempo	-2.907891	119.901788	Boulder	1
A	Lempo 5	Sesean Suloara'	Lempo	-2.901844	119.899759	Outcrop	1
A	Lempo 6	Bangkelekila'	To'yasa Akung	-2.902176	119.896037	Boulder	1
A	Lempo 7	Bangkelekila'	To'yasa Akung	-2.902145	119.896035	Boulder	1
A	Lempo 8	Sesean Suloara'	Sesean Matallo	-2.90563	119.890354	Boulder	4
A	Lempo 9	Sesean Suloara'	Sesean Matallo	-2.906104	119.886961	Boulder	1

(Continued)

Study area	Rock ID (cemetery)	District (Kecamatan)	Domain (Lembang/Kelurahan)	GPS location		Rock type	No. liang pa'
				Latitude	Longitude		
A	Lempo 10	Sesean Suloara'	Sesean Matallo	-2.907092	119.885974	Boulder	1
A	Lempo 11	Sesean Suloara'	Sesean Matallo	-2.906717	119.885598	Boulder	2
A	Lempo 12	Sesean Suloara'	Lempo	-2.908774	119.899015	Boulder	1
A	Lempo 13	Sesean Suloara'	Lempo	-2.908809	119.898664	Boulder	1
A	Lempo 14	Sesean Suloara'	Lempo	-2.907238	119.896848	Boulder	1
A	Lempo 15	Sesean Suloara'	Lempo	-2.907138	119.896711	Boulder	1
A	Lempo 16	Sesean Suloara'	Lempo	-2.907251	119.896545	Boulder	1
A	Lempo 17	Sesean Suloara'	Lempo	-2.90741	119.89621	Boulder	3
A	Lempo 18	Sesean Suloara'	Lempo	-2.907143	119.898004	Boulder	3
A	Lempo 19	Sesean Suloara'	Lempo	-2.905897	119.897402	Boulder	4
A	Lempo 20	Sesean Suloara'	Lempo	-2.905816	119.897476	Boulder	1
A	Lempo 21	Sesean Suloara'	Lempo	-2.90577	119.897503	Boulder	2
A	Lempo 22	Sesean Suloara'	Lempo	-2.905727	119.897404	Boulder	3
A	Lempo 23	Sesean Suloara'	Lempo	-2.905743	119.89734	Boulder	1
A	Lempo 24	Sesean Suloara'	Lempo	-2.90562	119.898054	Boulder	3
A	Lempo 25	Sesean Suloara'	Lempo	-2.905708	119.897869	Boulder	3
A	Lempo 26	Sesean Suloara'	Lempo	-2.909521	119.895076	Boulder	1
A	Lempo 27	Sesean Suloara'	Lempo	-2.909575	119.895194	Boulder	1
A	Lempo 28	Sesean Suloara'	Lempo	-2.909969	119.894698	Boulder	1
A	Lempo 29	Sesean Suloara'	Lempo	-2.910734	119.893966	Boulder	3
A	Lempo 30	Sesean Suloara'	Lempo	-2.910806	119.894097	Boulder	1
A	Lempo 31	Sesean Suloara'	Lempo	-2.911099	119.891876	Boulder	7

Study area	Rock ID (cemetery)	District (Kecamatan)	Domain (Lembang/Kelurahan)	GPS location		Rock type	No. liang pa'
				Latitude	Longitude		
A	Lempo 32	Sesean Suloara'	Lempo	-2.911813	119.8901	Boulder	1
A	Lempo 33	Sesean Suloara'	Lempo	-2.911618	119.888849	Boulder	1
A	Lempo 34	Sesean Suloara'	Lempo	-2.90654	119.89941	Boulder	1
A	Lempo 35	Sesean Suloara'	Lempo	-2.91112	119.89675	Boulder	1
A	Lempo 36	Sesean Suloara'	Lempo	-2.91171	119.89711	Boulder	1
A	Lempo 37	Sesean Suloara'	Lempo	-2.9117	119.89724	Boulder	1
A	Lempo 38	Sesean Suloara'	Lempo	-2.91159	119.89836	Boulder	1
A	Lempo 39	Sesean Suloara'	Lempo	-2.91252	119.8967	Boulder	1
A	Lempo 40	Sesean Suloara'	Sesean Matallo	-2.907591	119.885324	Boulder	5
A	Lempo 41	Sesean Suloara'	Sesean Matallo	-2.90769	119.885278	Boulder	1
A	Marante	Tondon	Tondon	-2.95397	119.93283	Cliff	3
A	Parinding 1	Sesean	Buntu Lobo	-2.927024	119.904523	Boulder	1
A	Parinding 2	Sesean	Buntu Lobo	-2.927024	119.904523	Boulder	1
A	Parinding 3	Sesean	Buntu Lobo	-2.927037	119.904429	Boulder	1
A	Parinding 4	Sesean	Buntu Lobo	-2.927128	119.903713	Boulder	17
A	Parinding 5	Sesean	Parinding	-2.927011	119.908023	Boulder	6
A	Parinding 6	Sesean	Parinding	-2.927003	119.907934	Boulder	1
A	Parinding 7	Sesean	Parinding	-2.926974	119.908334	Boulder	2
A	Parinding 8	Sesean	Parinding	-2.927502	119.907991	Boulder	2
A	Parinding 9	Sesean	Parinding	-2.927257	119.908789	Boulder	12
A	Parinding 10	Sesean	Parinding	-2.926981	119.908912	Boulder	7
A	Parinding 11	Sesean	Parinding	-2.927085	119.909443	Boulder	2

(Continued)

Study area	Rock ID (cemetery)	District (Kecamatan)	Domain (Lembang/ Kelurahan)	GPS location Latitude	GPS location Longitude	Rock type	No. liang pa'
A	Parinding 12	Sesean	Parinding	-2.926474	119.909215	Boulder	1
A	Parinding 13	Sesean	Parinding	-2.927187	119.90947	Boulder	4
A	Parinding 14	Sesean	Parinding	-2.927369	119.909461	Boulder	1
A	Parinding 15	Sesean	Parinding	-2.927433	119.909508	Boulder	3
A	Parinding 16	Sesean	Parinding	-2.927379	119.909567	Boulder	1
A	Tonga Riu 1/Lo'ko' Mata	Sesean Suloara'	Tonga Riu	-2.903986	119.862242	Boulder	95
A	Tonga Riu 2	Sesean Suloara'	Tonga Riu	-2.903993	119.862794	Boulder	3
A	Tonga Riu 3	Sesean Suloara'	Tonga Riu	-2.90406	119.862013	Boulder	1
A	Tonga Riu 4	Sesean Suloara'	Tonga Riu	-2.90406	119.862013	Boulder	2
A	Tonga Riu 5	Sesean Suloara'	Tonga Riu	Unknown	Unknown	Boulder	7
A	Suloara' 1/Pana'	Sesean Suloara'	Suloara'	-2.921052	119.870684	Cliff	32
A	Suloara' 2/Sele	Sesean Suloara'	Suloara'	-2.919884	119.871451	Boulder	28
A	Suloara' 3	Sesean Suloara'	Suloara'	-2.920048	119.871613	Boulder	2
A	Suloara' 4	Sesean Suloara'	Suloara'	-2.92458	119.87241	Boulder	2
A	Suloara' 5	Sesean Suloara'	Suloara'	-2.932601	119.868958	Boulder	1
A	Suloara' 6	Sesean Suloara'	Suloara'	-2.93228	119.868867	Boulder	4
A	Suloara' 7	Sesean Suloara'	Suloara'	-2.931608	119.867985	Boulder	3
A	Suloara' 8	Sesean Suloara'	Suloara'	-2.93153	119.867994	Boulder	1
A	Suloara' 9	Sesean Suloara'	Suloara'	-2.931459	119.867949	Boulder	1
A	Suloara' 10	Sesean Suloara'	Suloara'	-2.931367	119.86813	Boulder	3
A	Suloara' 11	Sesean Suloara'	Suloara'	-2.9313	119.868108	Boulder	4
A	Suloara' 12	Sesean Suloara'	Suloara'	-2.92874	119.872245	Boulder	1

Appendix

Study area	Rock ID (cemetery)	District (Kecamatan)	Domain (Lembang/ Kelurahan)	GPS location Latitude	GPS location Longitude	Rock type	No. liang pa'
A	Suloara' 13	Sesean Suloara'	Suloara'	-2.93076	119.87637	Boulder	5
A	Suloara' 14	Sesean Suloara'	Suloara'	-2.93091	119.87639	Boulder	1
A	Suloara' 15	Sesean Suloara'	Suloara'	-2.93095	119.87608	Boulder	2
A	Suloara' 16	Tikala	Embatau	-2.93248	119.87703	Boulder	1
A	Tallunglipu' 1/Tongka'	Tallunglipu	Tantanan Tallunglipu	-2.945873	119.907717	Cliff	11
B	Buntu Pune	Kesu'	Ba'tan	-2.98734	119.896128	Cliff	1
B	Ke'te' Kesu	Kesu'	Panta'nakan Lolo	-2.997835	119.910516	Cliff	4
B	La'bo' 1	Sanggalangi	La'bo'	-3.009446	119.916955	Cliff	2
B	La'bo' 2	Sanggalangi	La'bo'	-3.009569	119.916126	Cliff	1
Regency (Kabupaten): Tana Toraja							
B	Lemo 1	Makale Utara	Lemo	-3.042485	119.87732	Cliff	72
B	Lemo 2	Makale Utara	Lemo	-3.041732	119.876983	Cliff	8
B	Suaya	Sangalla	Bulian Massa'bu	-3.093786	119.902611	Cliff	6
B	Tampangallo 1	Sangalla	Tongko Sarapung	-3.0882	119.902666	Cliff	6
B	Tampangallo 2	Sangalla	Tongko Sarapung	-3.087466	119.902977	Cliff	4
C	Buri' 1/Batu Lappa'	Rembon	Buri'	-3.066683	119.789653	Cliff	14
C	Salu Liang	Malimbong Balepe'	Kole Sawangan	-3.075062	119.771755	Boulders	70
C	Ulin 1/Sanduni	Rembon	Ulin	-3.06333	119.795987	Sanduni	N/D

B. **List of erong cemeteries surveyed in June 2017**

Study area	Rock ID (cemetery)	Regency (Kabupaten)	District (Kecamatan)	Domain (Lembang/Kelurahan)	GPS location Latitude	Longitude
A	Lombok Parinding	Toraja Utara	Sesean	Buntu Lobo	-2.929088	119.905117
A	Marante	Toraja Utara	Tondon	Tondon	-2.953965	119.932828
A	Pana'	Toraja Utara	Sesean Suloara'	Suloara'	-2.921052	119.870684
A	Parinding 10	Toraja Utara	Sesean	Parinding	-2.926981	119.908912
A	Tongka'	Toraja Utara	Tallunglipu	Tantanan Tallunglipu	-2.945873	119.907717
B	Buntu Pune	Toraja Utara	Kesu'	Ba'tan	-2.98734	119.896128
B	Ke'te' Kesu	Toraja Utara	Kesu'	Panta'nakan Lolo	-2.997835	119.910516
B	Londa	Toraja Utara	Kesu'	Sangbua	-3.016059	119.876553
B	Pala'tokke	Toraja Utara	Sanggalangi	Paepalean	-3.011796	119.909601
B	Suaya	Tana Toraja	Sangalla	Bulian Massa'bu	-3.093786	119.902611
B	Tampangallo	Tana Toraja	Sangalla	Tongko Sarapung	-3.087956	119.902577

Bibliography

Adams, K.M. (2006) *Art as Politics: Re-crafting Identities, Tourism, and Power in Tana Toraja, Indonesia.* Honolulu, University of Hawai'i Press.

Adams, R.L. (2004) An ethnoarchaeological study of feasting in Sulawesi, Indonesia. *Journal of Anthropological Archaeology* 23(1), 56–78.

Adams, R.L. and Kusumawati, A. (2011) The social life of tombs in West Sumba, Indonesia. In R.L. Adams and S.M. King (eds) *Residential Burial: A Multiregional Exploration*, 17–32. Arlington VI, American Anthropological Association.

Adams, R. and Robin, G. (2022) Menhirs of Tana Toraja (Indonesia): A preliminary ethnoarchaeological assessment. In L. Laporte, J.-M. Large, L. Nespoulous, C. Scarre and T. Steimer-Herbet (eds) *Megaliths of the World,* Volume I, 307–321. Oxford, Archaeopress.

Beatty, A. (1992) *Society and Exchange in Nias.* Oxford, Clarendon Press.

Bigalke, T.W. (2005) *Tana Toraja: A Social History of an Indonesian People.* Singapore, Singapore University Press.

Bloch, M. (1971) *Placing the Dead: Tombs, Ancestral Villages and Kinship Organisation in Madagascar.* London, Seminar Press.

Bloch, M. and Parry, J. (1982) *Death and the Regeneration of Life.* Cambridge, Cambridge University Press.

Brisbois, E. and Douvier, F. (1980) *Les Toradja de Célèbes (Indonésie).* Paris, Hachète.

Buijs, K. (2006) *Powers of Blessing from the Wilderness and from Heaven: Structure and Transformations in the Religion of the Toraja in the Mamasa area of South Sulawesi.* Leiden, KITLV Press.

Crystal, E. (1985) The soul that is seen: The *tau tau* shadow of death, reflection of life in Toraja tradition. In J. Feldman (ed.) *The Eloquent Dead: Ancestral Sculpture of Indonesia and Southeast Asia*, 129–146. Los Angeles CA, UCLA Museum of Cultural History.

David, N. and Kramer, C. (2001) *Ethnoarchaelogy in Action.* Cambridge, Cambridge University Press.

Duli, A. (2014) Shape and chronology of wooden coffins in Mamasa, West Sulawesi, Indonesia. *TAWARIKH: International Journal for Historical Studies* 5(2), 117–186.

Duli, A. (2015) Typology and chronology of *erong* wooden coffins in Tana Toraja, South Celebes. *Time and Mind: The Journal of Archaeology, Consciousness, and Culture* 8(1), 3–20.

Duli, A. (2018) The roles of *liang* sites in the settlement system of the Torajan community. In M. Rohaizat Abdul Wahab, R. Mahwati Ahmad Zakaria, M. Hadrawi and Z. Ramli (eds) *Selected Topics on Archaeology, History and Culture in the Malay World*, 39–53. Singapore: Springer.

Duli, A., Rosmawati, M.N., Chia, S. and Ramli, Z. (2019) Archaeological study about burial tradition of Toraja ethnic, South Sulawesi, Indonesia. In S.S. Handoko, Aslinda and Sawirman (eds) *Proceeding of the 13th International Conference on Malaysia-Indonesia Relations (PAHMI). Contributions of Humanities and Social Sciences on the Direction of Malay Studies in the Era of Industry 4.0. August 21-24, 2019, Padang, West Sumatra, Indonesia*, 1–9. Berlin: Sciendo.

Fauvelle-Aymar, F.-X., Bruxelles, L., Mensan, R., Bosc-Tiessé, C., Derat, M.-L. and Fritsch, E. (2010) Rock-cut stratigraphy: Sequencing the Lalibela churches. *Antiquity* 84(326), 1135–1150.

Forth, G.L. (2001) *Dualism and Hierarchy: Process of Binary-Combination in Keo Society.* Oxford, Oxford University Press.

Fürer-Haimendorf, C. von (1939) *The Naked Nagas.* London, Methuen.

Gallay, A. (2006) *Les sociétés mégalithiques: Pouvoir des hommes, mémoire des morts.* Lausanne, Presses polytechniques et universitaires romandes.

Grubauer, A. (1913) *Unter Kopfjägern in Central-Celebes: Ethnologische Streifzüge in Südost- und Central-Celebes.* Leipzig, R. Voigtländers.

Hayden, B. (2016) *Feasting in Southeast Asia.* Honolulu, University of Hawai'i Press.

Hertz, R. (1907) Contribution à une étude sur la représentation collective de la mort. *L'Année Sociologique* 10, 48–137.

Hoskins, J.A. (1986) So my name shall live: Stone dragging and grave-building in Kodi, west Sumba. *Bijdragen tot de Taal-Land-en Volkenkunde* 142(1), 31–51.

Ikegami, S. (2018) The development of rock-cut tombs in the Japanese archipelago. *The Rissho International Journal of Academic Research in Culture and Society* 1, 171–199.

Ingold, T. (1993) The temporality of the landscape. *World Archaeology* 25(2), 152–174.

Ismail, A., Yusuf, A.M. and Safriadi (2019) Lamba tree: Environment wisdom and its resistance to development. *IOP Conference Series: Earth and Environmental Science* 343(1), 012094. https://iopscience.iop.org/article/10.1088/1755-1315/343/1/012094/pdf

Jannel, C. and Lontcho, F. (1992) *Les Toradjas d'Indonésie: Laissez venir ceux qui pleurent.* Paris, Armand Colin.

Janowski, M. (2020) Stones alive! An exploration of the relationship between humans and stones in Southeast Asia. *Bijdragen tot de -taal, -land en volkenkunde* 176(1), 105–146.

Jeunesse, C. and Denaire, A. (2018) Current collective graves in the Austronesian world: A few remarks about Sumba and Sulawesi (Indonesia). In A. Schmitt, S. Déderix and I. Crevecoeur (eds) *Gathered in Death: Archaeological and Ethnological Perspectives on Collective Burial and Social Organization*, 85–105. Louvain, Presses Universitaires de Louvain.

Jeunesse, C., Bec-Drelon, N., Boulestin, B. and Denaire, A. (2022) Aspects de la gestion des dolmens et des tombes collectives actuels dans les sociétés de l'île de Sumba (Indonésie). *Préhistoires Méditerranéennes* 9(2), 165–179.

Jeunesse, C., Le Roux, P. and Boulestin, P. (eds) (2016) *Mégalithisme vivants et passés: Approches croisées/Living and Past Megalithisms: Interwoven Approaches.* Oxford, Archaeopress.

Jong, E.B.P. de (2013) *Making a Living between Crises and Ceremonies in Tana Toraja: The Practice of Everyday Life of a South Sulawesi Highland Community in Indonesia.* Leiden, Brill.

Joyce, R.A. and Gillespie, S.D. (eds) (2000) *Beyond Kinship: Social and Material Reproduction in House Societies.* Philadelphia, University of Pennsylvania Press.

Kadang, K. (1960) *Ukiran Rumah Toradja [Toraja House Carvings].* Jakarta, Dinas Penerbitan Balai Pustaka.

Keane, W. (1997) *Signs of Recognition: Powers and Hazards of Representation in an Indonesian Society.* Berkeley CA, University of California Press.

Keers, W. (1939) Over de verschillende vormen van het bijzetten der doden bij de Sa'dan-Toradja. *Tijdschrift van het Koninklijk Nederlandsch Aardrijkskundig Genootschap* 56, 207–213.

Kis-Jovak, J.I., Nooy-Palm, H., Schefold, R. and Schilz-Dornburg, U. (1988) *Banua Toraja: Changing Patterns in Architecture and Symbolism among the Sa'dan Toraja of Sulawesi, Indonesia.* Amsterdam, Royal Tropical Institute.

Koubi, J. (1982) *Rambu solo', 'la fumée descend': le culte des morts chez les Toradja du Sud.* Paris, Éditions du CNRS.

Koubi, J. (2010) *Les Toradja et leur univers rituel.* Paris, CNRS Images.

Kruyt, A.C. (1924) De Toradja's van de Sa'dan, Masoepoe- en Mamasa-rivieren. *Tijdschrift voor Indische taal-, land-, en volkenkunde* 63, 81–175.

Lamesa, A., Hailay, Atsbha and Saint-Bézar, B. (2023) Églises rupestres du Təgray oriental et central: Résultats de prospections et hypothèses techniques et socio-économiques. *Annales d'Éthiopie* 25, 245–296.

Levi-Strauss, C. (1975) *La voie des masques.* Geneva, Albert Skira.

McKenzie, J. (1990) *The Architecture of Petra.* Oxford, Oxford University Press.

Melis, M.G. (ed.) (2000) *L'ipogeismo nel mediterraneo: Origini, sviluppo, quadri culturali. Atti del Congresso Internazionale (Sassari-Oristano, 23–28 maggio 1994).* Sassari, Università degli Studi di Sassari.

Muehlbauer, M. (2023) *Bastions of the Cross: Medieval Rock-cut Cruciform Churches of Tigray, Ethiopia*. Washington DC, Dumbarton Oaks Research Library and Collection.

Nobele, E.A.J. (1926) Memorie van overgave betreffende de onder-afdeeling Makale van den aftredende gezaghebber bij het binnenlandsch-bestuur, *Tijdschrift voor Indische taal-, land-, en volkenkunde* 66, 1–144.

Nooy-Palm, H. (1979) *The Sa'dan-Toraja: A Study of their Social Life and Religion. Vol. 1: Organization, Symbols and Beliefs*. The Hague, Martinus Nijhoff.

Nooy-Palm, H. (1986) *The Sa'dan-Toraja: A Study of their Social Life and Religion. Vol. 2: Rituals of the East and West*. Dordrecht, Foris.

Nooy-Palm, H. (1988) The Mamasa and Sa'dan Toraja of Sulawesi. In J.-P. Barbier and D. Newton (eds) *Islands and Ancestors: Indigenous Styles of Southeast Asia*, 86–105. Munich, Prestel.

Nooy-Palm, H. (1999) Sulawesi: The woodcarving of the Sa'dan and Mamasa Toraja. In D. Newton (ed.) *Art of the South Seas: Island Southeast Asia, Melanesia, Polynesia, Micronesia. The Collections of the Musée Barbier-Mueller*, 90–101. Munich, Prestel.

Nugraha, Y.S. (2019) Ethnomathematical Review of Toraja's Typical Carving Design in Geometry Transformation Learning. *Journal of Physics: Conference Series* 1280(4), 042020. Doi:10.1088/1742-6596/1280/4/042020

Parker Pearson, M. and Ramilisonina (1998) Stonehenge for the ancestors: The stones pass on the message. *Antiquity* 72(276), 308–326.

Parker Pearson, M., Godden, K., Ramilisonina, Retsihisatse, Schwenninger, J.-L., Heurtebize, G., Radimilahy, C. and Smith. H. (2010) *Pastoralists, Warriors and Colonists: The Archaeology of Southern Madagascar*. Oxford, British Archaeological Report S2139.

Polvé, M., Maury, R.C., Bellon, H., Rangin, C., Priadi, B., Yuwono, S., Joron, J.-L. and Soeria Atmadja, R. (1997) Magmatic evolution of Sulawesi (Indonesia): constraints on the Cenozoic geodynamic history of the Sundaland active margin. *Tectonophysics* 272, 69–72.

Rappoport, D. (2015) Sulawesi, 20 ans après. *Archipel* 89, 179–204.

Robin, G. (2017) What are bucrania doing in tombs? Art and agency in Neolithic Sardinia and traditional South-East Asia. *European Journal of Archaeology* 20(4), 603–635.

Sandarupa, S. (1997) *Life and Death in Toraja*. Ujung Padang, PT Torindo.

Sandarupa, S. (2016) 'The Voice of a Child': Constructing a moral community through retteng poetic argumentation in Toraja. *Archipel* 91, 231–258.

Sande, J.S. (1991) *Toraja in Carving's*. Makassar, Eigenverl.

Sevilla Casas, E. (2010) Shaft-and-chambers tombs in the necropolis of Tierradentro, Colombia. *International Journal of South American Archaeology* 6, 36–44.

Shanks, M. and Tilley, C. (1982) Ideology, symbolic power and ritual communication: a reinterpretation of Neolithic mortuary practices. In I. Hodder (ed.) *Symbolic and Structural Archaeology*, 129–154. Cambridge, Cambridge University Press.

Sherman, G.D. (1990) *Rice, Rupees, and Ritual: Economy and Society Among the Samosir Batak of Sumatra*. Stanford CA, Stanford University Press.

Steimer-Herbet, T. (2018) *Indonesian Megaliths: A Forgotten Cultural Heritage*. Oxford, Archaeopress.

Tsintjilonis, D. (2000) Death and the sacrifice of signs: 'Measuring' the dead in Tana Toraja. *Oceania* 71(1), 1–17.

Tsintjilonis, D. (2007) The death-bearing sense in Tana Toraja. *Ethnos Journal of Anthropology* 72(2), 173–194.

Veen, H. van der (1940) *Tae' (Zuid-Toradjasch)-Nederlandsch woordenboek met register Nederlandsch-Tae'*. The Hague, Nijhoff.

Veen, P. van der and Veen, M. van der (2023) *The Linguist's Family: Toraja in the Dutch East Indies, Prisoners of Wartime Japan, and the Republic of Indonesia*. Vancouver, Electromagnetic Print.

Volkman, T.A. (1979a) The arts of dying in Sulawesi. *Asia*, July/August 1979, 24–30.

Volkman, T.A. (1979b) The riches of the undertaker. *Indonesia* 28, 1–16.

Volkman, T.A. (1985) *Feasts of Honor: Ritual and Change in the Toraja Highlands*. Urbana IL, University of Illinois Press.

Waterson, R. (1986) The ideology and terminology of kinship among the Sa'dan Toraja. *Bijdragen tot Taal-, Land- en Volkenkunde* 142(1), 87–112.

Waterson, R. (1988) The house and the world: The symbolism of Sa'dan Toraja house carvings. *RES: Anthropology and Aesthetics* 15, 34–60.

Waterson, R. (1989) Hornbill, naga and cock in Sa'dan and Toraja woodcarving motifs. *Archipel* 38, 53–73.

Waterson, R. (1990) *The living House: An Anthropology of Architecture in South-East Asia*. Oxford, Oxford University Press.

Waterson, R. (1993) Taking the place of sorrow: The dynamics of mortuary rites among the Sa'dan Toraja. *Southeast Asian Journal of Social Science* 21(2), 73–96.

Waterson, R. (1995) Houses, graves and the limits of kinship groupings among the Sa'dan Toraja. *Bijdragen tot de Taal-, Land- en Volkenkunde* 151(2), 194–217.

Waterson, R. (1997) The contested landscapes of myth and history in Tana Toraja. In J.J. Fox (ed.) *The Poetic Power of Place: Comparative Perspectives on Austronesian Ideas of Locality*, 63–88. Canberra, Australian National University Press.

Waterson, R. (2009) *Paths and Rivers: Sa'dan Toraja Society in Transformation*. Leiden, KITLV Press.

White, L.T., Hall, R., Armstrong, R.A., Barber, A.J., BouDagher-Fadel, M., Baxter, A., Wakita, K., Manning, C. and Soesilo, J. (2017) The geological history of Latimojong region of western Sulawesi, Indonesia. *Journal of Asian Earth Sciences* 138, 72–91.

Wilcox, H. (1949) *White Stranger: Six Moons in Celebes*. London, Collins.